FOREWORD

The Workshop on *Measuring the Environmental Impacts of Agriculture*, hosted by the United Kingdom's Ministry of Agriculture, Food and Fisheries, was held in York on 22-25 September 1998, and included a one-day study visit to a large-scale farm enterprise and a hill farming area. The UK *Countryside Minister, Elliot Morley*, opened the Workshop, which brought together around 150 participants from agriculture and environment ministries in 26 OECD Member countries, the Commission of the European Communities, EUROSTAT, the European Environment Agency, the UN Environment Programme, FAO; and representatives from Birdlife International, the European Confederation of Agriculture; the International Federation of Agricultural Producers; the World Conservation Union; and the World-Wide Fund for Nature.

The Workshop was an integral part of the programme of work on agri-environmental policy issues in the OECD. Its key objective was to advance work on developing agri-environmental indicators, particularly where the work is less developed, by recommending appropriate indicators and methods of measurement for short- and medium-term development, and also share OECD countries' experiences and approaches in developing and using indicators for policy analysis. This publication is the result of work carried out by the OECD Joint Working Party of the Committee for Agriculture and the Environment Policy Committee. The two parent committees approved the present report in April 1999, and agreed that it be published under the responsibility of the OECD Secretary-General.

Part I of the Report sets the scene for the Workshop, with *Kevin Parris* (OECD Secretariat) describing the policy context of OECD's work in establishing a set of agri-environmental indicators (AEIs); followed by *David Pearce* (University of London) who places AEIs in the wider setting of measuring sustainable development. **Part II** provides a summary of the Workshop discussion and recommendations, with *Chris Doyle* (Scottish Agricultural College) providing an overall summary, followed by summary reports for the ten specific indicator areas examined at the Workshop. **Part III** examines AEIs as a tool for policy makers, and *Andrew Moxey* (University of Newcastle-upon-Tyne) analyses cross-cutting issues in developing indicators, *Paul Thomassin* (McGill University) explores the application of AEIs in policy analysis, and *David Baldock* (Institute for European Environmental Policy) reviews OECD country experiences in using AEIs for policy purposes. **Part IV** completes the Report with the text of the official statements at the Workshop. An **Annex** contains a list of the recommended OECD AEIs; the Workshop Agenda and contributors; and a list of participants.

The OECD expresses its appreciation to the UK authorities for the active role they played in hosting the Workshop, as well as to Austria, Canada, Finland, Japan, Norway, Spain, Sweden, Switzerland and the United States, for their financial contribution. OECD wishes to acknowledge the consultants and participants for their valuable input, and in particular, Steven Gleave, Nikolaj Bock and Gary Beckwith from the UK for helping to organise the event; the JSR Farming Group and North York Moors National Park Authorities for hosting the study visit; Jane Kynaston, the Conference Organiser; Françoise Bénicourt and Theresa Poincet, preparation of the Report; Véronique de Saint-Martin and Laetitia Reille for translation; and Richard Pearce for editing the text. The Report has also benefited from comments of OECD staff, notably: Wilfrid Legg, Kevin Parris, Seiichi Yokoi, Outi Honkutukia, Gérard Bonnis, Morvarid Bagherzadeh, Myriam Linster, Jeannie Richards and Jan Keppler.

OECD PROCEEDINGS

Environmental Indicators for Agriculture

Volume 2

Issues and Design
"The York Workshop"

ORGANISATION FOR ECONOMIC CO-OPERATION AND DEVELOPMENT

ORGANISATION FOR ECONOMIC CO-OPERATION AND DEVELOPMENT

Pursuant to Article 1 of the Convention signed in Paris on 14th December 1960, and which came into force on 30th September 1961, the Organisation for Economic Co-operation and Development (OECD) shall promote policies designed:

- to achieve the highest sustainable economic growth and employment and a rising standard of living in Member countries, while maintaining financial stability, and thus to contribute to the development of the world economy;
- to contribute to sound economic expansion in Member as well as non-member countries in the process of economic development; and
- to contribute to the expansion of world trade on a multilateral, non-discriminatory basis in accordance with international obligations.

The original Member countries of the OECD are Austria, Belgium, Canada, Denmark, France, Germany, Greece, Iceland, Ireland, Italy, Luxembourg, the Netherlands, Norway, Portugal, Spain, Sweden, Switzerland, Turkey, the United Kingdom and the United States. The following countries became Members subsequently through accession at the dates indicated hereafter: Japan (28th April 1964), Finland (28th January 1969), Australia (7th June 1971), New Zealand (29th May 1973), Mexico (18th May 1994), the Czech Republic (21st December 1995), Hungary (7th May 1996), Poland (22nd November 1996) and Korea (12th December 1996). The Commission of the European Communities takes part in the work of the OECD (Article 13 of the OECD Convention).

OECD CENTRE FOR CO-OPERATION WITH NON-MEMBERS

The OECD Centre for Co-operation with Non-Members (CCNM) was established in January 1998 when the OECD's Centre for Co-operation with the Economies in Transition (CCET) was merged with the Liaison and Co-ordination Unit (LCU). The CCNM, in combining the functions of these two entities, serves as the focal point for the development and pursuit of co-operation between the OECD and non-member economies.

The CCNM manages thematic and country programmes. The thematic programmes, which are multi-country in focus, are linked to the core generic work areas of the Organisation (such as trade and investment, taxation, labour market and social policies, environment). The Emerging Market Economy Forum (EMEF) and the Transition Economy Programme (TEP) provide the framework for activities under the thematic programmes. The EMEF is a flexible forum in which non-members are invited to participate depending on the theme under discussion. The TEP is focused exclusively on transition economies. Regional/Country programmes, providing more focused dialogue and assistance, are now in place for the Baltic countries, Brazil, Bulgaria, China, Romania, Russia, the Slovak Republic (a candidate for accession to the OECD), and Slovenia.

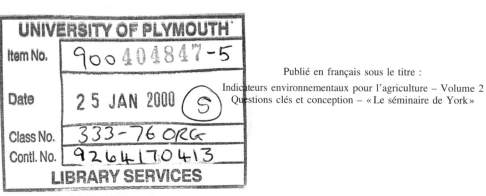
Publié en français sous le titre :
Indicateurs environnementaux pour l'agriculture – Volume 2
Questions clés et conception – « Le séminaire de York »

TABLE OF CONTENTS

Part I:

POLICY CONTEXT OF OECD AGRI-ENVIRONMENTAL INDICATORS

Kevin Parris,
Policies and Environment Division,
Directorate for Food, Agriculture and Fisheries, OECD, Paris, France

MEASURING SUSTAINABLE DEVELOPMENT: IMPLICATIONS FOR AGRI-ENVIRONMENTAL INDICATORS

David Pearce,
Centre for Social and Economic Research on the Global Environment (CSERGE),
University College, London and University of East Anglia, United Kingdom

USING AGRI-ENVIRONMENTAL INDICATORS TO ASSESS ENVIRONMENTAL PERFORMANCE

Paul J. Thomassin
McGill University, Quebec, Canada

DEVELOPING AND USING AGRI-ENVIRONMENTAL INDICATORS FOR POLICY PURPOSES: OECD COUNTRY EXPERIENCES

David Baldock
Institute for European Environmental Policy, London, United Kingdom

Part IV:
Official Statements

Tables

Annex Tables

Figures

Boxes

PART I:

SETTING THE SCENE

POLICY CONTEXT OF OECD AGRI-ENVIRONMENTAL INDICATORS

by
Kevin Parris,
Policies and Environment Division,
Directorate for Food, Agriculture and Fisheries, OECD, Paris, France[1]

1.   Introduction

OECD Member countries have given a high priority to developing a set of agri-environmental indicators (AEIs), and this paper gives the context to AEI work by addressing the following questions:

- Why is OECD developing a set of indicators?

- What are the agri-environmental policy issues being addressed by indicators?

- What is the role of indicators in agri-environmental policy decision-making?

- How is the OECD work being developed?

- How are indicators informing the policy dialogue and what are the future challenges?

2.   Why is OECD developing a set of indicators?

The issues of sustainable agriculture, the environment and natural resource use are high on both the domestic and international policy agenda. This is illustrated by the agricultural policy reform programmes underway in many OECD Member countries (partly as a result of the Uruguay Round Agreement on Agriculture), which in most cases include policies to promote sustainable agriculture and address environmental and natural resource concerns. In addition, recently a number of international environmental agreements with important implications for agriculture have been signed, including the United Nations Conference on Environment and Development (UNCED) Rio

---

1.   The author wishes to thank his colleagues Wilfrid Legg, Seiichi Yokoi, Outi Honkatukia, Gérard Bonnis, Morvarid Bagherzadeh, Myriam Linster, Jan Keppler and Jeannie Richards for their helpful comments in preparing this paper, and also thank Françoise Bénicourt and Jane Kynaston for their preparation of the paper.

Declaration and Agenda 21, the Convention on Biological Diversity, and the Kyoto Protocol commitments to reduce greenhouse gas emissions.

The heightened importance of agricultural and environmental policy issues and the need for better information concerning sustainable development has been highlighted at three recent OECD Ministerial meetings. *The meeting of OECD Agricultural Ministers,* 5-6 March 1998 (OECD, 1998a):

- *Outlined a set of shared goals,* including that the agro-food sector "contributes to the sustainable management of natural resources and the quality of the environment".

- *Adopted a set of policy principles,* including the need for countries to "take actions to ensure the protection of the environment and sustainable management of natural resources in agriculture by encouraging good farming practices, and create the conditions so that farmers take both environmental costs and benefits from agriculture into account in their decisions".

- *Identified a role for OECD* to, amongst other tasks, "foster sustainable development through analysing and measuring the effects on the environment of domestic agricultural and agri-environmental policies and trade measures".

The meeting of OECD Environment Ministers, 2-3 April 1998 (OECD, 1998b):

- *Stressed the need for* "deepening work on integrating environmental concerns into key economic sectors, such as agriculture and fisheries, transport, energy, and into trade, investment and fiscal policy, at Secretariat level and through co-operative arrangements among OECD committees".

- *Recommended OECD should* "further develop and adopt a comprehensive set of robust indicators to measure progress toward sustainable development, in concert with sustainable development initiatives of other international agencies, to be used in country reviews and outlook reports, including the second cycle of environmental performance reviews".

- *Endorsed* "the principles set out in the OECD Council Recommendation on Environmental Information" (OECD, 1998c). The principles encourage OECD countries to intensify their efforts to upgrade the extent and quality of environmental data, indicators and information dissemination systems. This is to support the preparation and implementation of result-oriented strategies and effective policies concerning the environment and sustainable development.

The meeting of the OECD Council at Ministerial Level, 27-28 April 1998 (1998d):

- *"Agreed that the achievement of sustainable development is a key priority for OECD countries.* They encouraged the elaboration of the Organisation's strategy for wide-ranging efforts over the next two years in the areas of climate change, technological development, sustainability indicators, and the environmental impact of subsidies".

- *"Asked the OECD to enhance its dialogue with non-Member countries in these areas* and to engage them more actively, including through shared analyses and development of strategies for implementing sustainable development".

It is against this background that *work on sustainable development is now a major horizontal activity for the OECD*, with a report planned for the OECD Ministerial Council Meeting in 2001, which would also be an input into the Rio + 10 Conference in 2002 (OECD, 1998e). The report will provide a policy strategy to help achieve sustainable development with emphasis on the economic, social and environmental dimensions. It will draw on all relevant OECD work including agriculture and AEIs, and will initially focus on four projects: climate change; support measures, taxes and resource pricing; technology; and indicators of sustainable development.

The main objectives of the OECD project on indicators of sustainable development, are to:

- Review progress toward establishing a common framework for the development of sustainable development indicators.

- Explore how progress can be made on technical aspects of indicator development, such as physical and monetary measures and spatial scales.

- Advance work on an integrated and practical set of indicators for policy analysis, including monitoring and evaluation.

3. What are the agri-environmental policy issues being addressed by indicators?

3.1 Sustainable agriculture

The basic long-term challenge facing agriculture is to produce sufficient food and industrial crops efficiently, profitably and safely, to meet a growing world demand without degrading natural resources and the environment (OECD, 1995). While improvements in agricultural productivity have been substantial they have often been accompanied by resource degradation, such as soil erosion and water depletion (OECD, 1998f), and damage, for example, to genetic diversity, which in some cases has impaired growth in farm output. But farmers have also made positive contributions to landscapes and the maintenance of rural communities. Agricultural land can also provide important habitats for wildlife and act as a sink for greenhouse gases.

Because of differences in climate, agro-ecosystems, population density and levels of economic development, the relative importance of particular environmental issues varies widely from one OECD country to another and within countries. These differences are also reflected in perceptions across and within OECD countries as to what is meant by the "environment" in agriculture. For some, the "environment" covers only biophysical and ecological aspects. For others, landscape, cultural features, and rural development are also important. In recent years, the quality and safety of food, and the welfare of farm animals, have become more prominent policy issues, perceived as being closely related with the environment.

Different perceptions regarding what constitutes the "environment" in agriculture influence the distinction between the "harmful" and the "beneficial" environmental effects of agriculture. This is linked to the role ascribed to "agriculture", and the "jointness" of agricultural activities and environmental outcomes. In other words, agriculture is often perceived not only as a "crop and livestock" production activity, but also as an activity which is closely linked to the provision of environmental services.

The environmental implications of farmers' actions, however, are not always incorporated in their costs and revenues, such as when agricultural chemicals leach into groundwater, thereby raising the costs of treating water for drinking (OECD, 1997a). Or, for example, when farmers refrain from taking actions that would add to environmental services, such as conserving land as habitat for wildlife, because effective ways of making the beneficiaries compensate farmers for the associated costs are lacking.

From a policy perspective a distinction needs to be made between those agricultural activities that benefit, and those that harm the environment, and those activities that are accounted, or not accounted for by farmers in their decisions (OECD, 1997b). Whichever "baseline" is chosen, the direction of change of an environmental effect will indicate whether there has been an improvement or deterioration in environmental performance. This requires quantitative information, including indicators.

3.2   Agricultural policies, policy reform and the environment

Given the long period of **government intervention in agriculture**, in most OECD countries, there is a particular interest in the environmental effects of agricultural policy reform. Agricultural policy reform to increase the market orientation of the sector and liberalise agricultural trade, while taking account of the multiple roles played by agriculture, was agreed by the OECD Ministerial Council in 1987, and most recently in 1998 by OECD Agriculture Ministers. Reform efforts are currently underway, in part to meet the commitments made by countries under the Uruguay Round Agreement on Agriculture.

The cost of agricultural support in the period 1986-88 accounted for 1.7 per cent of GDP in the OECD area, and agricultural support was equivalent to about 41 per cent of the value of agricultural production, as measured by the OECD's Producer Subsidy Equivalent (PSE). Overall support to agriculture has since fallen and, for the period 1996-98, was equivalent to 33 per cent of the value of OECD agricultural production, while the share of overall support in GDP was 1.3 per cent. But the average figures for OECD disguise a wide range of developments among countries and commodities (OECD, 1999a).

In many cases environmental problems have been aggravated by agricultural and trade policies that distort price signals by linking support to agricultural commodities, or by disguising the costs of agricultural inputs. The economic distortions created by such policies can lead to environmentally inappropriate patterns and location of production, environmentally harmful use of inputs, and discourage the development and adoption of farming technologies less stressful on the environment (OECD, 1998g).

Over the last decade, an important component of the wider package of *agricultural policy reforms in all OECD countries* has been the introduction of new programmes addressing environmental issues. Agri-environmental policy measures in some countries seek to reduce agricultural pollution, maintain the landscape, and promote biological diversity and rural development, often through payments to farmers. All OECD countries have passed regulations to deal with agricultural problems related to pollution of air and water, and damage to ecologically sensitive areas, and many have set codes of "good" farming practices, and linked existing agricultural support programmes to environmental conditions.

Some countries place great stress on community-based approaches to agricultural resource management, which aim to motivate farmers to take greater responsibility for the local management of natural resources (OECD, 1998h). Research and development into technologies and farming methods that are less environmentally stressful and promote sustainable agriculture are also being encouraged by most OECD countries. The dissemination of this research and new technologies through agricultural extension services is also playing an important role in reducing risks to the environment.

The reform of agricultural policies should improve the domestic and international allocation of resources, reduce incentives to the over-use of polluting chemical inputs and to farm environmentally sensitive land. In other words, through reducing output and input use (due to a combination of lower output prices and changes in relative factor prices), the reforms would tend to reverse the harmful environmental impacts associated with commodity and input specific policy measures. But in those cases where agricultural policies are associated with maintaining farming activities that provide environmental benefits, policy reform can reduce environmental performance (OECD, 1998i).

As agricultural policy reforms have been introduced only recently in most countries, some caution needs to be exercised when assessing the overall environmental effects. There are few examples of fundamental policy reforms, and the degree of decoupling of support for commodity production is limited. Environmental effects are complex, often site-specific, take time to become evident, and the data and other evidence required to quantify environmental effects are not always available. Also many other developments have been occurring at the same time, which may or may not be linked to policy reform, such as changes in: technology, consumer preferences, environmental pressure group demands, environmental regulations, industrial pollution, climate change, and the relative prices of farm inputs (including chemicals) and outputs.

OECD work has recognised that agricultural policy reform is a necessary, but not always a sufficient condition to improve the environmental performance of agriculture. Given the diversity and site specificity of agro-ecological conditions, local, farmer-based approaches, coupled with relevant research, development, training, information and advice would appear to be high on the list of "sound" policy practices. These approaches focus on the "public good" aspects of agriculture, reflect the differences across farming, and allow for the development of market-based innovations.

4.   What is the role of indicators in agri-environmental policy decision-making?

The OECD has set itself three key *objectives for developing AEIs* (OECD, 1997c) to meet the needs of policy makers and other stakeholders:

- *Contribute to the available information* on the current status and trends in the condition of the environment and natural resources in agriculture.

- *Improve the understanding* of agri-environmental processes and the impact of agricultural policies on the environment.

- *Provide a tool* to monitor and evaluate agricultural and environmental policies to help improve policy effectiveness in promoting sustainable agriculture.

The development of environmental indicators is relatively recent compared to work on economic and social indicators. But whereas the latter are often concerned with the monetary measurement of human phenomena, environmental indicators aim to capture the relationship between the biophysical "natural" environment and human activities, usually measured in physical terms. This, in part, explains why environmental and sustainable agriculture indicators present a greater challenge, especially with respect to the spatial and temporal dimensions, and the issues of linkages and valuation.

Space: the range of scales to measure AEIs varies from the field, farm, watershed, through to the ecozone and national levels. The capability to develop and measure indicators for a range of spatial scales is constrained by: the ability to extrapolate data from the field/farm level to higher levels; the trade-offs that occur with gains in coverage at higher levels but loss of the detail/variation at lower scales; and that information at different scales may require different indicators depending on the use, and users, of the information. These vary from farmers to national level policy makers (McRae *et al.*, 1995). From the OECD perspective, data need to be captured at as detailed a level as possible then aggregated to the national level with some expression of the variation around the national average indicator value.

Time: the variations in the time scales of different environmental effects of agriculture range from the short term, such as the impact on wildlife from pesticide use; medium term, for example depletion of groundwater reserves; and to the long term, which may involve decades in the case of soil erosion or centuries in the case of climate change. The impacts on the environment from agricultural policies, economic and societal pressures may also have different time lags and consequences. While this problem is not uncommon to socio-economic indicators, there is nonetheless an important difference, as a key focus of sustainable development is intergenerational concerns. Most indicators, however, use a time series approach showing current trends and, as noted by David Pearce in this report, this presents a key challenge for indicator construction in terms of the current-future trade-off.

Linkages: the sustainable development concept emphasises the links between the economic, social and environmental dimensions (Rennings and Wiggering, 1997). The OECD indicators recognise these dimensions of sustainable agriculture, for example, through farm financial (economic); socio-cultural (social) and water quality (environmental) indicators. But it is also necessary to show the linkages between the three dimensions of sustainable agriculture, for example, between measures of resource productivity and the health consequences of agri-environmental impacts. Balancing economic imperatives (e.g. food production), with environmental impacts (e.g. conserving landscapes) and social concerns (e.g. preserving rural communities), requires some means of weighing up these impacts/concerns, such as the use of cost-benefit frameworks.

Valuation: the use of a cost-benefit framework highlights the need to develop AEIs that use a common monetary unit rather than physical measures, so that trade-offs and priorities can be more easily gauged by policy makers and the public. As noted in the many of the papers presented to this Workshop, efforts are underway to develop methods of valuing environmental costs and benefits.

While there is a gap between the current development of AEIs (indicator supply) and expectations for indicator delivery by policy makers and other stakeholders (indicator demand), AEIs are essential to make well-informed policy choices. Without them there is a risk of making short-sighted and flawed decisions (Ervin *et al.*, 1995).

Figure 1. Criteria for selecting environmental indicators

POLICY RELEVANCE AND USEFULNESS FOR THE USER	An environmental indicator should:
	♦ provide a representative picture of environmental conditions, pressures or society's responses;
	♦ be simple, easy to interpret and able to show trends over time;
	♦ be responsive to changes in the environment and related human activities;
	♦ provide a basis for international comparisons;
	♦ be either national in scope or applicable to regional issues of national significance;
	♦ have a threshold or reference value against which to compare it, so that users are able to assess the significance of the values associated with it.
ANALYTICAL SOUNDNESS	An environmental indicator should:
	♦ be theoretically well founded in technical and scientific terms;
	♦ be based on international standards and international consensus about its validity;
	♦ lend itself to being linked to economic models, forecasting and information systems.
MEASURABILITY	The data required to support the indicator should be:
	♦ readily available or made available at a reasonable cost/benefit ratio;
	♦ adequately documented and of known quality;
	♦ updated at regular intervals in accordance with reliable procedures.

*Extract from OECD (1993), *Environmental Indicators for Environmental Performance Reviews,* Paris, France.
**These criteria describe the "ideal" indicator and not all of them will be met in practice.

In order to develop appropriate indicators, work undertaken by OECD has suggested that these should possess a number of attributes. These have been derived from previous OECD work on AEIs (OECD, 1993; and 1997c). They are outlined in Figure 1, and imply that indicators must be:

- policy-relevant — they should be demand (issue) rather than supply (data) driven, and address the key environmental issues faced by governments and other stakeholders in the agriculture sector;

- analytically sound — based on sound science, but recognising that their development involves successive stages of improvement;

- easy to interpret — they should communicate essential information to policy makers;

- measurable — they need to be feasible in terms of current or planned data availability and be cost effective in terms of data collection, processing and dissemination.

5.   How is the OECD work being developed?

5.1   *Establishing a framework to develop AEIs*

The OECD has established the ***Driving Force–State–Response (DSR) framework*** through which to develop indicators (Figure 2). This framework addresses a set of questions related to the linkages between causes, effects and actions.

- What is causing environmental conditions in agriculture to change, for example, changes in farm chemical input use (***Driving forces***)?

- What are the effects of agriculture on the environment, for example, the impacts on soil, water, air, and natural habitats (***State***)?

- What actions are being taken to respond to the changes in the state of the environment, for example, by farmers, consumers, the food industry and governments, such as promoting sustainable agriculture by community based approaches (***Responses***)?

Analysis of the linkages between the causes and effects of agriculture's impact on the environment is a key element to better guide policy makers in their responses to changes in environmental conditions in agriculture. At present, however, these linkages are not yet fully developed. Further work needs to be undertaken on the linkages between indicators in the DSR framework, in order that causal relationships and feedbacks may be better understood and more easily expressed by policy makers and other stakeholders (Hodge, 1997; and Andrew Moxey, David Pearce and Paul Thomassin in this report). In addition, the framework does not takes into account the effects on agriculture which result from changing environmental conditions, such as industrial pollution and climate change.

Examination of agri-environmental linkages in the DSR framework highlights the need to further:

- Develop understanding of the physical, chemical and biological factors that relate variations in agricultural practices, input use and production to changes in environmental quality.

- Improve knowledge of the economic, socio-cultural and policy factors that determine and influence the effects of agricultural activities on the environment.

- Quantify each component in the DSR framework, through collection and verification of analytically sound data.

20

Figure 2. The Driving Force-State-Response Framework to address agri-environmental linkages and sustainable agriculture

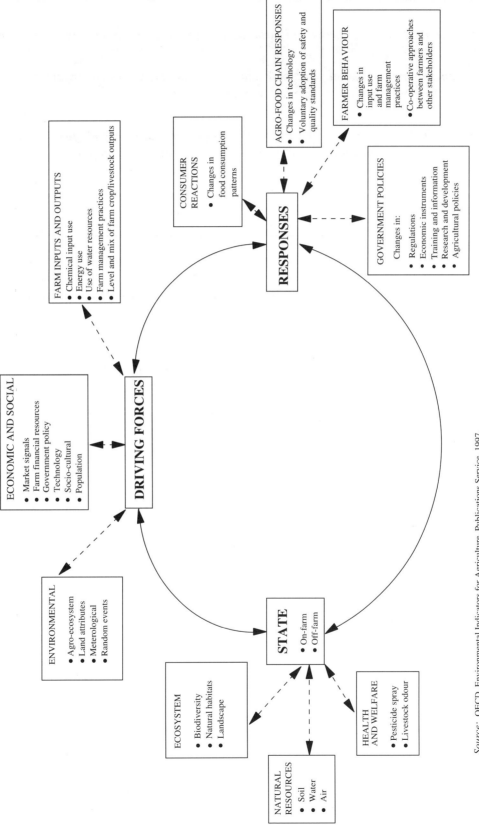

Source: OECD, Environmental Indicators for Agriculture, Publications Service, 1997.

21

Within the context of the DSR framework the OECD has currently identified a number of priority areas for which indicators are being developed to cover primary agriculture's:

- *use of natural resources and farm inputs:* nutrients, pesticides, water and land;

- *environmental impact on:* soil and water quality, land conservation; greenhouse gas emissions, biodiversity, habitats and landscape; and,

- *interaction between environmental, economic and social factors,* such as farm management practices; farm financial resources; and socio-cultural characteristics.

A key objective for the York Workshop is to develop a core set of indicators for those areas which are least advanced in terms of their conceptual and analytical basis and methods of measurement. These are: water quality; water use; soil quality; land conservation; biodiversity; wildlife habitats, landscape, farm management, farm financial resources, and socio-cultural issues. The indicators covering nutrient use, pesticide use and greenhouse gases, are already in progress, with OECD work at a relatively advanced stage for nutrient balances (OECD, 1999b), pesticide use data and also agricultural greenhouse gas emission/sink information, while pesticide risk indicators are being developed in the OECD Pesticide Forum (OECD, forthcoming).

5.2   *Developing the indicators and integrating them into policy analysis*

Work on developing the indicators essentially involves a four-stage process (the example of the nitrogen balance indicator is used to illustrate this process, see OECD, 1999b), as outlined below.

- *Developing the conceptual and analytical understanding*, of the various agricultural and environmental processes and interactions underpinning each of the agri-environmental areas. In the case of nitrogen use this has involved incorporating knowledge of nitrogen cycles and nitrogen input-output relationships in agriculture.

- *Identifying the indicators and methods of measurement*, that are representative of each area and are easily understood. The soil surface nitrogen balance, for example, is the starting point to develop indicators for nutrient use, and future work is intended to explore farm gate nutrient balances, as well as develop a more integrated approach covering the linkages between farm nutrient management practices, soil and water quality.

- *Collecting data and calculating the indicators*, once consensus has been reached among OECD Member countries on the relevant indicators and methods of measurement. For the nitrogen soil surface balances, for example, data have now been collected and processed in co-operation with OECD countries for the period 1985 to 1997 (OECD, 1999b).

- *Integrating the indicators into policy analysis*, through analytical tools which foster better understanding of trends in environmental performance and monitor and evaluate the impact of policies on the environment in agriculture. The preliminary national estimates of the nitrogen soil surface balances (expressed as kilograms of nitrogen surplus per hectare of the total agricultural area, OECD, 1999b) have already been used as supporting information in a number of recent OECD reports (OECD, 1998i; 1998j; 1998k; and 1999a).

5.3   Co-ordinating the work with OECD Member countries and international organisations

The co-ordination of AEI work with OECD Member countries, involves drawing on the expertise and special interest of countries across agri-environmental areas. This approach, in harnessing the experience already built up by many Member countries, is regarded as a successful feature of OECD's work (see David Baldock in this report). In addition, the new OECD Member countries —the Czech Republic, Hungary, Korea, Mexico and Poland — are making a valuable contribution to the AEI work, across a range of different areas including on nutrient balances, land conservation and biodiversity. An exchange of information on the OECD work on AEIs is also beginning to develop between the OECD Secretariat and some non-Member countries, such as Argentina and Slovenia. Also as AEI work is completed and published it could be of value to a wider audience, especially many developing countries.

Within the OECD Secretariat and its various working groups, the Joint Working Party on Agriculture and the Environment (JWP) provides the main forum to discuss and reach consensus on developing AEIs. The JWP has also played an important role in integrating the AEIs into its policy-related work (OECD, 1998g; 1998i; and 1998j). The AEIs are also beginning to be incorporated into various activities of the Committee for Agriculture (COAG, see OECD, 1998k), including the annual report, *Monitoring and Evaluation Report of Agricultural Policies in OECD Countries* (OECD, 1999a).

In the Environment Policy Committee (EPOC) the work on AEIs is closely related to and co-ordinated with the activities of the Group on the State of the Environment (SOE). The SOE Group has been developing several types of indicators including the OECD Core Set of environmental indicators (OECD, 1998l), various sets of indicators for the integration of environmental concerns into sectoral policies (e.g. transport, OECD, 1999c; energy, OECD, 1993b) and indicators derived from natural resource and other environmental accounts (OECD, 1993a; 1997d; 1998l). AEI work is also feeding into the series of Environmental Performance Reviews in OECD countries, with a number of Reviews including a special focus on agriculture. Work is also being co-ordinated with the EPOC Forum on Climate Change and Energy, the Greenhouse Gas Inventory Group together with related work on energy balances by the OECD International Energy Agency.

Work is also being planned to investigate the feasibility of incorporating AEIs into various quantitative studies of the COAG, including the OECD model for medium term agricultural commodity forecasts (AGLINK) to evaluate the market and trade impacts of policy options to meet environmental standards (e.g. taxes, subsidies and regulation). These include reducing greenhouse gas emissions from agriculture, as well as the environmental effects of different scenarios for agricultural production and trade. In addition, future work will examine how the environmental effects of different agricultural policy measures might be analysed, drawing on the agri-environmental work, as part of the COAG work on developing a Policy Evaluation Matrix.

The AEI work also contributes to other OECD activities related to sustainable development, including the OECD-wide activity to develop a set of Sustainable Development Indicators, as previously discussed in the second section of this paper. The OECD Group of the Council on Rural Development is working to establish a set of rural territorial indicators, which will include an environmental component. This work will be co-ordinated with that on AEIs (Meyer, 1998).

There are also a considerable number of international organisations, both governmental (IGOs) and non-governmental (NGOs), which have already begun to develop framework and indicators related to agriculture and the environment, with most of the key players present at this Workshop. In developing the AEIs OECD has co-ordinated its work closely with that of the European Union's statistical agency, EUROSTAT, most recently in areas of work on agricultural nutrient balances and landscape.

OECD is also engaged in an active exchange of information related to AEIs with various IGOs including the UN Food and Agriculture Organisation (FAO); European Environment Agency (EEA); World Bank; UN Environment Programme (UNEP); UN Framework Convention on Climate Change (UNFCCC), and some NGOs active in the area, namely Birdlife International, the International Federation of Agricultural Producers (IFAP) and the World-Wide Fund for Nature (WWF).

6.   How are indicators informing the policy dialogue and what are the future challenges?

Indicators are providing valuable information by revealing where a problem may be emerging that might require a policy response, and as a contribution to monitoring the environmental effects of actions taken by farmers in response to changing policy incentives or disincentives. This work is already starting to quantify environmental trends in agriculture, and provide consistent time series on key environmental parameters. The input into this Workshop will provide a significant impetus to the work.

From the work already undertaken, both in OECD Member countries and internationally, it is clear that such *indicators are helping to better inform the policy decision-making process by*:

- Showing the linkages between agricultural policies and environmental performance, and between the economic, social and environmental dimensions of sustainable agriculture.

- Quantifying the choices between agriculture and environmental outcomes.

- Providing an additional tool to help monitor and evaluate agricultural policy developments.

- Helping to identify the main agri-environmental issues which require attention, so contributing to policy efforts to improve environmental performance.

- Revealing which policy measures are most effective in achieving agri-environmental objectives.

- Throwing light on the positive and negative environmental externalities that arise from agricultural activities, such as landscape and pollution of groundwater.

- Facilitating comparisons across countries within a standard methodology.

There is, however, still the need to further advance the OECD work to better inform the future policy dialogue. This will involve *a number of challenges in the future development of indicators to*:

- Improve the availability, timeliness and quality of data to calculate indicators, especially aspects concerning spatial and temporal coverage, and also agricultural land use.

- Integrate the indicators into quantitative analysis, including the valuation of the environmental effects of agricultural activities, especially to examine the impacts on the environment of alternative policy measures and the consequences of possible agricultural production and trade projections on the environment.

- Widen the dialogue for developing indicators, not only with policy makers in OECD Member countries, but also with non-Member countries and those concerned with agriculture and the environment. In particular, to promote dialogue with farmers, environmental interest groups, scientists and researchers, consumers and the agro-food industry.

Recent OECD Ministerial meetings have expressed the need to provide better information and indicators regarding the effects on the environment of domestic agricultural and agri-environmental policies and trade measures. This work is also necessary as an important building block in measuring progress toward sustainable agriculture and more broadly, sustainable development.

The essential challenge is to develop policy-relevant, measurable, comparative and easily understood environmental indicators for agriculture. An analogy might be the canaries once used by miners to alert them to the danger of the build-up of underground gases. This simple indicator, namely the collapse of the canary, was to alert miners that they needed to escape from the mine very quickly. But policy makers and society at large cannot escape from having to make decisions and choices concerning the environment in agriculture. Indicators can be a valuable tool to better inform the policy decision-making process, and this Workshop offers an excellent opportunity to meet that need.

BIBLIOGRAPHY

ERVIN, D., S. BATIE and M. LIVINGSTON (1995), "Developing Indicators For Environmental Sustainability: The Nuts and Bolts: Introduction and Symposium Summary", pp. 1-7, in S. Batie, (ed.), *Developing Indicators For Environmental Sustainability: The Nuts and Bolts*, Special Report (SR) 89, Proceedings of the Resource Policy Consortium Symposium, Washington, D.C.

HODGE, T. (1997), "Toward a Conceptual Framework for Assessing Progress Toward Sustainability", *Social Indicators Research*, Vol. 40, pp. 5-98.

McRAE, T., *et al.* (1995), "Role and Nature of Environmental Indicators in Canadian Agricultural Policy Development", pp. 117-42, in Batie, ed., *op cit.*

MEYER, von H. (1998), "The Insights of Territorial Indicators", *OECD Observer*, No. 210, February/March, Paris.

OECD (1993a), *OECD Core Set of Indicators for Environmental Performance Reviews: a Synthesis Report by the Group on the State of the Environment*, Environment Monograph No. 83, Paris.

OECD (1993b), *Indicators for the Integration of Environmental Concerns into Energy Policies*, Environment Monograph No. 97, Paris.

OECD (1995), *Sustainable Agriculture — Concepts, Issues and Policies in OECD Countries*, Paris.

OECD (1997a), *Agriculture, Pesticides and the Environment: Policy Options*, Paris.

OECD (1997b), *The Environmental Benefits of Agriculture — The Helsinki Seminar*, Paris.

OECD (1997c), *Environmental Indicators for Agriculture*, Paris.

OECD (1997d), *OECD Environmental Data — Compendium, 1997 Edition*, Paris.

OECD (1998a), Meeting of the Committee for Agriculture at Ministerial Level, OECD *News Release*, 6 March, Paris.

OECD (1998b), "OECD Environment Ministers' Shared Goals for Action", *OECD News Release*, 3 April, Paris.

OECD (1998c), *Recommendation of the Council on Environmental Information*, Paris.

OECD (1998d), "OECD Council Meeting at Ministerial Level", *OECD News Release*, 28 April, Paris.

OECD (1998e), *OECD Work on Sustainable Development*, a discussion paper on work to be undertaken over the period 1998-2001, see OECD web-site: http://www.oecd.org.

OECD (1998f), *Sustainable Water Management in Agriculture — The Athens Workshop*, Paris.

OECD (1998g), *The Environmental Effects of Reforming Agricultural Policies*, Paris.

OECD (1998h), *Co-operative Approaches to Sustainable Agriculture*, Paris.

OECD (1998i), *Agriculture and the Environment: Issues and Policies*, Paris.

OECD (1998j), *Agricultural Policies in OECD Countries: Monitoring and Evaluation 1998*, Paris.

OECD (1998k), *Agricultural Policy Reform: Stocktaking of Achievements*, a discussion paper for the Meeting of the OECD Committee for Agriculture at Ministerial Level, March, Paris.

OECD (1998l), *Towards Sustainable Development — Environmental Indicators*, Paris.

OECD (1999a), *Agricultural Policies in OECD Countries: Monitoring and Evaluation 1999*, Paris.

OECD (1999b), *OECD National Soil Surface Nitrogen Balances: Preliminary Estimates 1985-1997*, Paris.

OECD (1999c), *Indicators for the Integration of Environmental Concerns into Transport Policies*, Environment Monograph No. 80, Paris.

OECD (forthcoming, 2000), *Agri-environmental Indicators: Methods and Results,* Paris.

RENNINGS, K. and H. WIGGERING, (1997), "Steps Towards Indicators of Sustainable Development: Linking Economic and Ecological Concepts, *Ecological Economics*, Vol. 20, No. 1, January, pp. 25-36.

MEASURING SUSTAINABLE DEVELOPMENT: IMPLICATIONS FOR AGRI-ENVIRONMENTAL INDICATORS

by
David Pearce
Centre for Social and Economic Research on the Global Environment (CSERGE),
University College, London and University of East Anglia, United Kingdom

1.   Introduction

It is a privilege to be asked to present this keynote address to such an important conference. I hope that what I have to say will have some bearing on the excellent papers that follow.

I shall do my best to "tell a story" about sustainable development, about its measurement, and about the implications for the way we monitor the complex interactions between agriculture and the environment. Good stories begin with questions, proceed to analysis, and end with answers to the questions. Let us begin with the questions and also suggest immediate broad answers before going on to look at them in a little more detail.

2.   The questions about sustainable development

2.1   What is sustainable development?

First, and inevitably, we are compelled to ask *what is sustainable development*? While this question has spawned a huge and sometimes agonising literature, it is really a straightforward question. Sustainable development is development that lasts, and which is not therefore threatened by actions we take now and which will have their major consequence in the future.

We could no doubt deliberate for ever about how far into the future we should look when deciding how long we want to guarantee this development. We have examples of actions taken which are designed to protect many, many generations to come: the Kyoto Protocol to the Framework Convention on Climate Change at least has this intent, as does the Montreal Protocol and the various Amendments, all concerned to reduce the depletion of the stratospheric ozone layer. The various attempts to design disposal for long-lived radioactive waste have time horizons of at least thousands of years, perhaps hundreds of thousands of years. But the reality is that the further we look into the future the less sure we can be what will happen, least of all in a world of such rapidly changing technology as we have now. Seeking guarantees for the very distant future is therefore probably a hopeless task and, as we shall see, it may also be a counterproductive one because of the risk that it will divert attention from actions which should benefit the least advantaged now. One way round this time horizon problem is to develop rules which commit any one generation to care for, say, the next two or three generations. The subsequent generations therefore inherit that commitment and look forwards to the next few generations. This sequencing of obligations avoids any one generation having to worry about what happens 1000 years hence, or even 150 years hence. It also imparts a flexibility to those concerns because the obligations will change as generations change in their tastes and moralities. Finally, we do not have to worry too much about what future generations want — an often cited problem for the pursuit of sustainable development — because the immediate next generations are here already and we can talk to them. They are our children and grandchildren.

That leaves us with the problem of *defining development*. I propose not to spend much time on that question, not because it is not interesting, but because I suggest that what we have to do to secure it will be, by and large, invariant with how we define development. Moreover, if we follow the sequencing rule, there is flexibility to adapt to differing definitions of development as time goes by. For now, let us define development as rising per capita well-being of human populations. This is a working definition: for example, it omits the well-being of other sentient creatures and I, for one, regard that as very important. But including their well-being is a topic for another time, although, of course, their well-being should figure in our agri-environmental indicators.

2.2   *Why measure sustainable development?*

Why do we want to measure sustainable development?

There is a strong force for devising indicators.[2] If we genuinely embrace sustainable development, we must have some idea if the *path* we are on is heading towards it or away from it. There is no way we can know that unless we know what it is we are trying to achieve — i.e. what sustainable development means — and unless we have indicators that tell us whether we are on or off a sustainable development path.

2.3   *The state of the indicator business*

While these observations about the *goal* and the *path to the goal* are extremely simple, they automatically rule out many so-called "sustainability indicators". The indicator business, if we can call it that, has largely developed on the basis of what there is rather than what there should be. Information has often been produced which is in search of a purpose. That is not to say the indicators are not interesting: very often they are. But they are very often not indicators of sustainability. In other cases we have indicators which are unquestionably useful as inputs to sustainability indicators. Take as an example the generally excellent set of indicators produced by the United Kingdom (Department of Environment, UK, 1996), and consider just two examples. In the "land cover and landscape" section we have data on pesticide use in the UK, and in a different section on "wildlife and habitats" we have some data on changes in bird populations. But not only are the two not linked in any way in the Department of Environment Report, but neither section in the Report even mentions that there is probably a link. Similarly, the section on "soil" selects three indicators of soil quality —

---

2.   It is as well to be aware that some people do *not* want measures of sustainability. Indicators might show them up in a bad light, in which case it is always better to say that sustainable development is a fuzzy concept and has many meanings, but is, of course, something we all support. If indicators develop in particular ways, they may also force decision makers to address questions they prefer not to address, for example, the real, underlying causes of environmental degradation rather than the cosmetic causes which can be addressed and for which, perhaps with adequate "spin doctoring", good publicity can be obtained.

concentrations of organic matter, acidity and nutrient concentrations in topsoils — but does not mention erosion, despite there being very detailed erosion data for much of the UK. As a general observation, what is missing in these data sets are best guesses at dose-response functions — the link between emissions or environmental insults and damage — and any measure of just how important these impacts are. Of course, the indicator business is developing and we may see these measures in the future. For the moment, however, what we have are environmental indicators, not sustainability indicators. We return to these points later.

Obviously, since sustainable development is a "good thing", its definition is easily hijacked to serve the purposes of almost any advocacy. But if we keep to the meaning of the words, then it is difficult to quarrel with the interpretation that *sustainable development comprises increases in real per capita well-being over time*. In other words, the very first indicator we require is one which measures that well-being. There are, of course, many such indicators, from the crudest, which is Gross National Product (GNP) per capita, modified measures of "green" National Product, to reduced form composite indicators such as the UNDP's Human Development Index (HDI), along with many that are, unfortunately, spurious.[3] I propose not to say anything more about these, save that they are measures of what has happened rather than what will happen. They are *ex post* rather than *ex ante*. As such, they do not tell us much about the future, and, as we saw, the essence of sustainable development is that it is future oriented. This in turn suggests that we should focus not only on the indicator of per capita well-being, but on the *conditions for achieving sustained increases in those indicators*.

2.4   How do we get sustainable development?

The basic theory of how to achieve sustainable development is well developed in the economics literature and is conveniently summarised in Atkinson *et al.* (1997). I will offer no more than the sketchiest outline. For future generations to be better off in developmental terms than we are today they must have the *capacity* to generate more well-being than we have.[4] Indeed, as there are going to be many more people in the future, that increase in capacity must be quite marked if per capita well-being is to improve. But on what does well-being depend? It depends on the capability for self-realisation and fulfilment and we know that this depends heavily on education, skills and knowledge. Let us call that *human capital*.

We know that the capacity to generate high per capita output of goods and service, upon which well-being undeniably depends, is determined by the availability of human capital and also stocks of machinery and infrastructure, or *man-made capital*.

---

3.   The most notable being the "Index of Sustainable Economic Welfare", as advanced by Daly and Cobb (1989) for the United States, and Jackson and Marks (1994) for the United Kingdom. For a penetrating critique, see Atkinson (1995).

4.   Notice that the focus is on capability rather than achievement. If future generations fail to make the best of what we leave them, that is their fault, not ours.

It is also widely believed that capacities to generate well-being depend on the set of stable relationships between individuals and groups within society, on such things as trust and keeping one's promises. This is *social capital*. In the agricultural context we need to include in social capital the preservation of rural communities and the rural "way of life", although, as we shall see, that need not equate with some of what happens in modern day agriculture. Actually *measuring* social capital is perhaps the major challenge that faces the practical implementation of sustainable development.[5]

Finally, we have now come to recognise that the stock of environmental assets, or *natural capital*, is important for well-being, not just because they create amenity and beauty, but because they affect our physical and mental health as well.

So, the future capacity to sustain development depends on these stock of capital. And this gives us the clue to getting sustainable development. As a general rule, these stock of capital should not decline through time: we should pass on to the next generation at least as much capital as we have to day. More precisely, per capita stocks should not decline through time, a rule that has come to be known as the *constant capital rule*.[6] There is one way that we can modify this statement. An existing stock of capital can do more "work", i.e. provide more well-being, if it embodies the latest technology. So, our constant capital rule could be restated as keeping a technology-weighted per capita index of total capital at least constant through time. Yet another way of thinking about it to say that total capital stocks should be constant or rising, and that technological change should grow at least as fast, and preferably faster, than population change. This formulation comes very close to the way economists formulated sustainable economic growth requirements in the 1970s (e.g. Stiglitz, 1979).

2.5   *Indicators for sustainable development*

Now we can begin to see what kinds of indicators we need: they are indicators of the stock of capital assets — man-made, natural, social and human. In terms of environmental indicators, we see that what we require are measures of stocks and changes in stocks. To some extent, we have this, for example in indicators of the stock of fish in fishery, and in indicators of environmental degradation. The former is the stock, the second is a measure of the net change in that stock. But capital assets are aggregations of many individual assets, so we cannot say much about the overall stock unless we have aggregate measures. The only exception would be if *everything* was increasing or decreasing, and that is unlikely. There are many different ways of aggregating assets but, strictly, we need a weighting procedure that applies across all forms of capital. The reason for this is that we have not advanced a case for supposing that any one form of capital matters more than any other. So, not only do we need to add up the different component parts that make up, say, natural capital, but we need in principle to be able to add up natural, human, social and man-made capital. The only measuring rod so far devised that enables us to do that, is money. That is, of course, exactly how, say, a company's assets are measured, i.e. in terms of their money value. We should not be too surprised to discover that this solution at the company level becomes the same solution at the level of a nation, or a region, or even the world.

---

5.   For a discussion see World Bank (1997), and for an approach based on measuring the willingness to pay for charitable concerns see Foster *et al.* (1999).

6.   For extensive discussion and analysis of this rule, see Atkinson *et al.* (1997).

The transition from definition of sustainable development to spelling out the conditions for achieving it is probably not very controversial. The idea that money is the appropriate measuring rod probably is controversial. I propose not to dwell on the justification for money as the means of aggregating capital stocks since I have set this out many times in other publications (e.g. Pearce, 1993). The essential point, however, is that what money is doing is acting as the measuring rod of price. In turn price reflects individuals' willingness to pay for something. And willingness to pay reflects people's preferences. So, money becomes the means of measuring preferences. And the preferences that matter are those of the ordinary person.

From this briefest of discussions we see that not only can we define sustainable development, not only can we say what has to be done to get it, but we also have an approach to measuring sustainable development based on measures of capital stocks and changes in those stocks. Our indicator is still not totally forward-looking and I doubt if any indicator ever will be. But what we have learned is that persistent running down of capital assets is not consistent with sustainable development, *even if* at one at the same time some indicator of well-being is rising. *Put another way, GNP, say, a Human Development Indicator could well be increasing but the underlying path of development is unsustainable.* Such situations are not only possible, they are not unusual, as my colleagues and I have shown on a number of occasions (see, for example, Atkinson *et al.*, 1997). We know it colloquially as "selling the family silver" or "living off capital". There is therefore some comfort in learning that our approach has a basis in common good sense.

Any measure of well-being is therefore a measure of "development" only in so far as development ignores the capital basis for progress. The focus should be on the underlying capital base. Monitoring this will not guarantee sustainability, but it will help.

3.   **Is sustainable development focused too much on future generations?**

Along with most discussions of sustainable development we have focused on what we have to do to secure *future* development, i.e. the well-being of future generations. But this exclusive focus is not justified either in terms of the early discussion of sustainable development (e.g. the Brundtland Commission, see World Commission on Environment and Development, 1987), or the reality.

Intergenerational equity concerns the costs that the current generation may impose on future generations. Examples of such "intertemporal externalities" include any cumulative burden of pollution — arising from, for example, climate change, stratospheric ozone depletion and persistent use of chemicals and heavy metals — and irreversible loss of biological diversity. These impacts may reduce future generations' well-being. There are three main reasons why these intergenerational effects are thought to be especially important:

1. future generations have no "voice", i.e. they appear not to be represented in terms of decision-making today. As such, they cannot put pressure on current generations to reduce these externalities;

2. there is no mechanism for compensating future generations for the damage inflicted on them by a current generation;

3. the effects that are imposed could be irreversible, meaning that future generations cannot avoid the damages imposed on them, however rich they are.

We have already discussed the *first issue* and have concluded that, at the very least, it is not as serious as might be thought.

As to the *second issue*, there is in fact a mechanism for compensating future generations: economic growth. If current generations ensure that future generations are better off in income terms, then current generations have effectively sacrificed something (their expenditure on consumption) in order to save resources which are then invested to create wealth for the future. This raises an important issue: should we spend existing resources to reduce environmental burdens on people 100 years hence, or should we spend the same resources increasing the income of those same people? Moreover, if we devote resources to reducing future environmental burdens, what is the cost of that policy in terms of foregone income increases for the poor now? As Schelling (1998) has argued, there can be no case for giving priority to the future poor over the present poor. Indeed, there is a strong case for considering the present poor to have priority: the descendants of the present poor are likely to be richer than the present poor. And, just to make the ethical issue more complex, what if helping the poor now to raise their incomes we impose a cost on future generations because the policies of stimulating economic growth now cause environmental pollution in the future?

The issue is one of a trade-off that is not easily resolved. Unless obligations to future generations are somehow intrinsically different to obligations to present generations, there is a strong case for focusing on the poor of the current generation, and that focus may be inconsistent with policies designed to help the future.

There may be a case for giving priority to worrying about intergenerational equity if the costs imposed by the present on the future are irreversible. For then we are committing *all* future generations to suffer the consequences. Again, this *third issue* is not straightforward. There are a great many irreversibilities in decision-making and we create those irreversibilities every day. Amenity, forests, wetlands, coastal resources are lost all the time. In principle those losses are reversible. In practice it is difficult to see how they can be reversed against the backdrop of an unavoidable 50 per cent increase in the world's population and perhaps a 100 per cent increase in 100 years.[7] Irreversibility *per se* cannot be a reason for focusing solely on future generations. Of course, irreversibility is not irrelevant. It matters because the loss of assets is equivalent to a loss of options and a loss of information. All options — i.e. all freedoms to choose — and all information have an economic value, and that economic value is being lost through irreversible decisions. Factoring in this kind of loss is therefore important for decision-making.

Presumably what matters more is the *scale* of the irreversible effect. Suppose climatic change gives rise to an increase in diseases associated with changed patterns of rainfall and temperature. If the effect is not easily reversed, then the current generation has imposed a cost on future generations. But income growth tends to be associated with health improvement (Pearce, 1998), so that policies designed to control climate change could result in increases in diseases simply because some income growth is sacrificed. The difficulty is to determine whether the increased disease from climate control is greater or less than the increased disease from climate change.[8] Pursuing intergenerational equity is potentially consistent with improving the health of future generations at the expense of the health of current generations.

---

7.   The 50 per cent increase is unavoidable because of "population momentum". See World Bank (1994).

8.   This is "risk-risk" or "health-health" analysis. See e.g. Viscusi (1994).

34

There is a challenge here for indicator construction. Casual inspection of indicator documents shows that they focus very much on the time-series approach, i.e. on whether a chosen indicator is going up or down over time. There is scant acknowledgement of the current-future trade-off. Indeed, there is hardly a mention of the *social distribution* of environmental degradation or environmental benefits, i.e. who gains and who loses. Work of this kind has advanced in North America under the banner of "environmental justice", but it barely exists in Europe and the rest of the world. Here, then, is a further major challenge to practitioners of indicator construction.

4. Indicators of sustainable development

The preceding analysis automatically leads to indicators of sustainable development. Since my colleagues and I have surveyed the development of these indicators elsewhere (Atkinson *et al.*, 1997; Hamilton *et al.*, 1996; and Pearce *et al.*, 1998), I offer only the briefest of checklists here.

4.1 Economic indicators

Wealth

From the constant capital rule it follows immediately that a measure of *total wealth* over time is a capital-based indicator of sustainable development. In turn this means that each component of wealth must be measured in commensurate units to secure aggregation, and money is the obvious (and probably the only) measuring rod here. The World Bank (1997; see also Kunte *et al.*, 1998) has published the first estimates of total wealth for 1994, a single year. The measure excludes social capital which, we noted, presents formidable challenges for measurement, but includes man-made, natural and human capital. The main finding is that human capital dominates the wealth of advanced economies. Since there is, as yet, no time series we can say little about changes over time and therefore little about sustainability. Nonetheless, this approach has high promise for the future.

"Green" National Product

Probably the greatest effort has gone into modified GNP measures to account for changes in capital stocks that are either excluded from conventional accounts or which are included in a manner that makes the change in their stocks difficult to identify. Reviews of these estimates can be found in Atkinson *et al.* (1997), Hamilton and Lutz (1996) and Hamilton *et al.* (1994). While the debate about the "exact" measure of green net national product (gNNP) continues, consensus seems to form around a measure of the following nature:

$$gNNP = C + I - r(R\text{-}g) - p(e\text{-}d)$$

where:
	gNNP	=	modified net national product
	C	=	consumption
	I	=	investment, including appreciation of human capital (education)
	r	=	rental on natural resources
	R	=	harvest or extraction of the natural resource
	g	=	regeneration rate for the natural resource (g=0 for non-renewables)
	p	=	(marginal) willingness to pay to avoid pollution (the "price" of pollution)
	e	=	emissions of pollutants
	d	=	rate of natural degradation of pollutants (rate of assimilation).

Notice that while gNNP appears to be a refined measure of "well-being" rather than of the underlying change in capital stocks, the changes in capital stocks are accounted for in the expressions for resource change, pollution and investment. One problem with gNNP is that it is difficult to interpret as an indicator of sustainability. While it is typically below conventional measures of NNP does this mean the economy is not sustainable? It does not tell us because, of course, the gNNP measure could be rising over time. Another way of thinking about the problem is that gNNP has no "benchmark" for sustainability, no value below which we would say the economy is unsustainable and above which it is sustainable. This problem has led to the development of the "genuine savings" concept.

"Genuine savings"

Pearce and Atkinson (1993) suggested a reformulation of the gNNP constructs to derive an indicator of sustainability which does have a "natural" origin: genuine savings. The resulting indicator has been refined and is estimated for over 100 countries in World Bank (1997). The savings indicator (Sg) is:

$$Sg = S - r(R\text{-}g) - p(e\text{-}d)$$

Where the notation is as for gNNP, but Sg is genuine savings and S is gross savings.[9] The similarities with the gNNP formulation are obvious, but the attraction is that, as a general proposition, Sg < 0 indicates non-sustainability and Sg > 0 indicates sustainability. There is also a neat intuition to it since the rule becomes "save more than the depreciation on your assets" which is commonplace to any businessman.

The other attraction of the genuine savings formulation is that it has the potential for surprise. An economy can have all the outward appearances of success, but be revealed by the genuine savings indicator to be unsustainable or nearly so. Pearce and Atkinson (1993) suggested this was the case for the United Kingdom throughout much of the 1980s, whilst Sg indicators for the United States show a barely sustainable economy due to low savings rates.

Weak and strong sustainability

All the economic indicators described are indicators of *"weak sustainability"*. What this means is that they rely on an aggregation of different forms of capital, so that, by assumption, all forms of capital can be substituted for each other. It is therefore quite consistent with this form of sustainability for natural environments to be degraded so long as other forms of capital are built up (or vice versa). This is not an attractive concept to many with strong environmental preferences, and hence there is a strong movement to adopt *"strong sustainability"* measures in which one or other form of capital has its stock constrained to be non-declining. Essentially what is being argued is that the environment has no substitutes, a notion introduced into the debate in Pearce *et al.* (1989). There is a fairly virulent debate between advocates of the strong and weak approach. Issues arising are:

---

9.   One could equally put I in place of S; see Pearce *et al.* (1998).

1. the extent to which natural assets really are substitutable, an issue often confused by the failure to distinguish changes at the margin and the fairly obvious non-substitutability of the total "stock" of the environment;[10] and

2. the extent to which, if there is non-substitutability, natural capital is the only asset which manifests this attribute. It is arguable that societies are more likely to be non-sustainable due to social breakdown, i.e. social capital is non-substitutable. In any event, strong sustainability also requires weak sustainability: it makes no sense to prevent, say, natural capital from declining if the overall capital stock declines.

4.2   Ecological indicators

Parallel to the economic literature there has been an "ecological" literature on sustainability. This has been directed more explicitly at agricultural systems.

Two broad forms of indicator have been developed. The first is based on the traditional notion of *carrying capacity* and the second is based on the notion of *diversity and resilience*.

Carrying capacity measures

In principle, carrying capacity is a simple notion. First, find an indicator of the maximum flow of a given resource such that the stock does not decline over time, i.e. the maximum sustainable yield. Divide this by the minimum intake of the resource per person for survival, and the result is the maximum sustainable population. The resources in question have typically been expressed in food terms but energy and water may be just as important. If actual populations exceed the maximum sustainable population, then the situation is unsustainable.

The notion is attractive because it is fairly simple to calculate. However, it diverges from the economic definitions which are embedded in the concept of a rising level of per capita well-being. As expressed above, carrying capacity is not a sustainable development indicator, but a "survival" indicator. It tells us how many people can survive, and promises nothing about their future well-being. Indeed, it is easily misconstrued as a policy objective rather than as a maximum constraint. Carrying capacity is also problematic because of the ability to trade in most cases: a fuelwood shortage might be alleviated by importing fuelwood and exporting food, for example.

Perhaps more helpful are indicators that focus in on the capital stock and look at competition for that stock. Probably the most famous of these is an index of *net primary product appropriation* (Vitousek *et al.*, 1986; Haberl, 1997). Net primary production (NPP) is the total amount of energy produced by photosynthesis minus what is required by plants themselves for their own life processes. NPP is diverted by humans when they convert resources to their own purposes, reducing the amount of energy available for other species. Vitousek *et al.* (1986) estimated that while only 2 per cent of ocean NPP was appropriated by humankind, some 40 per cent of land-based NPP was appropriated. The appropriation of global NPP was some 25 per cent. If such measures can be projected over time,

---

10.   An example of such confusion is the much publicised paper by Costanza *et al.* (1997). For a critique, see Pearce (1998).

for example by linking appropriation rates to population growth, then it should be possible to say more about the rate of erosion of the natural capital base. Since the process of conversion is also the main reason for the loss of biological diversity, projections of species loss are also possible (Vitousek *et al.*, 1997; Ehrlich and Ehrlich, 1997).

Diversity-resilience measures

The second approach to ecological measures of sustainability rests on the notion that diverse ecological systems are more resistant to stress and shock than less diverse systems (Holling, 1973). Here *resilience* becomes the performance (or "output") measure of sustainability and this might be measured by the variability of economic activity. Thus, the amplitude of cycles about trends would be a first measure of resilience, with increasing amplitude over time being a warning sign that the system is becoming less resilient. Such measures have in fact been detected for agricultural systems: world and regional grain output, for example, exhibits exactly this form of behaviour (Hazell, 1984; Anderson and Hazell, 1989). Similar findings apply to the "green revolution": crop yields have become higher but more variable. The sources of this increased variability are many but undoubtedly include the increasing "homogenisation" of crop varieties and even agricultural technologies. Increasing variability might not matter if variations are always reversible. Some question whether this is the case, however, because of the discontinuities and thresholds that they believe mark ecological functioning. Thus, a given major downward section of the cycle might cross a threshold from which there is no recovery.

The link between homogenisation and variability explains the focus on diversity as the underlying, or "input" based, indicator of sustainability. Here the idea would be to extend the already rich array of measures of diversity in ecosystems to whole sectors, such as agriculture, and perhaps beyond to the workings of the entire economy (Perrings and Common, 1997). As yet, an easy-to-interpret empirical literature based on these ideas has not appeared, and this offers another challenge to researchers. As Atkinson *et al.* (1997) note as well, there is a problem with diversity as a "capital-based" indicator because of the absence of a benchmark "zero" below which there is unsustainability and above which there is sustainability.

5.   Sustainable development and the agricultural sector

Armed with the overview of a theory of sustainable development we can now turn more explicitly to the agricultural sector and the kinds of indicators we need. The indicators emerge from three basic questions: how large do we want the agricultural sector to be? what composition of output should the sector produce? and how should the sector produce that output?

5.1   How big should the agricultural sector be?

While this may appear a strange question to ask, it is of fundamental importance. Table 1 shows some crude indicators of existing size. The notable contrast is between the relative unimportance of the sector in advanced economies when expressed in terms of GDP or labour force, but the continuing high importance of the sector in terms of land use. Of course, part of the reason for the continuing high proportion of land devoted to agriculture in rich countries is the level of output (and input) subsidisation, so it is difficult to say what these proportions would be if a real open market was in place. The scale of land use also serves as a first indicator of the sustainability problem: any activity taking up over a third of land area is at risk of creating widespread environmental problems.

Returning to the question of the appropriate size, the economic answer is fairly straightforward. The market should decide on the appropriate scale *provided all the environmental and social costs of agricultural activity are taken into account*. Put another way, once all economic distortions are removed, the resulting size of the sector is "optimal". There are three major distortions: (a) subsidies to inputs and outputs, (b) environmental and social costs such as pollution, greenhouse gases, soil erosion and biodiversity loss, and (c) environmental *benefits* from the creation of amenity and a rural "way of life".

Table 1. Size of the agricultural sector

Region	Agriculture value added as % of GDP (1995)	% of total labour force in agriculture (1990)	% of total land area under crops and pasture (1994)
Low-income countries excluding China and India	33	67	40
Low-income countries including China and India	25	69	44
Middle-income countries	11	32	33
High-income countries	2	5	36

Source: World Bank, *World Development Report 1997*, Oxford University Press, Oxford, various tables.

All this suggests that agri-environmental indicators should include explicit treatment of these three factors. The use of the market as a guiding criterion for determining size simply reflects the earlier emphasis on preferences and willingness to pay. Markets are, in essence, mechanisms for revealing preferences. Of course, those preferences are weighted by the incomes of those expressing their willingness to pay. Given the focus of sustainable development on equity concerns, we may want to consider introducing equity explicitly into the indicators. One problem here is that this might imply that altering the size of the sector to serve equity goals is a good thing, when it might be better to correct the equity problem through progressive taxation and redistributive measures and let the market adjust.

5.2   What should the agricultural sector produce?

The answer to the question about the *composition* of the output of the agricultural sector is similar. The market should determine what is produced provided all distortions are removed. To see that such a policy would alter what is consumed consider the case of water. Some crops are water intensive and water is, in many countries, notoriously under-priced (Xie, 1996). The effect of correctly pricing water to reflect true opportunity costs — i.e. what the water would produce in an alternative use — would certainly change the composition of agricultural output in many countries, especially those where there is water scarcity, as in some Mediterranean countries. Even in Eastern England, there is evidence to suggest that the current use of water for irrigated crops is economically and environmentally unjustified (Bate and Dubourg, 1995).

5.3   How should the agricultural sector produce its output?

Here the answer is that producers of output should decide on the choice of technologies for production, but once again subject to the internalisation of any environmental and social costs they generate.

5.4   Agri-environmental indicators

The answers to all three questions — size, output composition and technologies for production — all require that we understand the external environmental impacts from agriculture. This in turn provides us with the rationale for the selection of indicators. We need to focus on the impacts that agriculture has. But even here the problem is complex. Any external effect is a composite of several things: a "dose" or "insult" to the environment, a "response" in the form of reduced biodiversity, human health impairment etc.; the "stock at risk"; and a "valuation" of that response, i.e. something that tells us whether that response is important or not.

Dose-response functions

The first major task is to investigate just what the relationship is between agricultural activity and environmental impacts, i.e. the "dose-response function". It is here that indicators are often weakest. Take the issue of the marked decline in bird populations in the United Kingdom. This is related to a complex of factors but changing farm practices, hedgerow removal, different crop choices and the use of pesticides are all implicated (Campbell *et al.*, 1997). But the relationship is not easy to quantify or, to put it another way, we do not know the value of the dose-response function between, say, pesticides and changing bird populations. Indicators which juxtapose pesticide use and bird population declines are therefore suggestive rather than conclusive. Much the same goes for pesticides use and farmers' health in countries where there is clear abuse of pesticide applications. Somewhat more surprisingly, we have strong data on water erosion of soils in Europe and good data on crop production and its economic value, yet we know little about the true economic costs of soil erosion in Europe.[11]

Stock at risk

A second requirement is to know how many people or ecosystems or species are at risk from agricultural impacts. Yet the stock at risk is not known with much precision in many cases.

Valuation

The third requirement is some indicator of importance. We have suggested willingness to pay as a powerful indicator. When it comes to agricultural "externalities", however, we find that we do not have that much research to guide us on willingness to pay. Smith (1992) has assembled data for United States agriculture but with the aim of stimulating further research since, even for the United States, the data do not permit a very reliable estimation of the "true" sectoral contribution to overall human well-being. Other attempts to estimate externalities have been interesting but flawed — see,

---

11.   Indeed, we seem to know more about those costs in developing countries. Arguably it is more important to know the costs in, say, Mali or Malawi, but it is strange that we can only hazard guesses at the cost in Europe when the data are so much better.

for example, Pimentel *et al.* (1992) and Steiner *et al.* (1995) on US pesticide damages (for a critique, see Pearce and Tinch, 1998). Again, there is a huge research agenda here. Accounting for positive and negative externalities from agriculture amounts to a recognition of the multiple functions of agriculture — i.e. as a provider of food and fibre, as a source of biodiversity, and as a provider of the rural way of life (Swiss Federal Office of Agriculture, 1998).

5.5   *Valuing the good things*

Lastly, the indicator literature has tended to focus on "bad" environmental impacts, such as pollution. Of course, indicators of bad things can go down or up, i.e. the trend can be good or bad, but little attention appears to be being paid to impacts which are themselves good. There are two ways in which good impacts could be handled.

The first is obvious: provided food reaches those who need it most, more agricultural output is a good thing. One might therefore accompany all agri-environmental indicators with data that remind us of this.

The second good thing is not so obvious. While we focus on pesticides and fertilizers, pollution incidents and run-off, it is as well to remember that many of these things contribute to higher agricultural productivity. And without that extra productivity on existing land, the race to convert existing non-agricultural land, especially forest land, would be even faster than it is. We should therefore remember the counterfactual, hard as it may be to identify and measure it, i.e. the land area we do *not* convert because of increased agricultural productivity. That foregone land conversion buys us precious time as we struggle to make agriculture and the environment more compatible in the search for sustainable development.

6.   Conclusions

The "indicator business" is now well established and occupies substantial research and policy time. But much of the business lacks focus, mainly because the indicators have been data-generated rather than problem-generated. I mean by this that indicators have been developed because the data are there. Some valuable efforts have been made to put those indicators into a conceptual framework such as the driving force, (or pressure), state, response framework. Even here, however, there has been limited appreciation of what the real driving forces are that give rise to threats to sustainable development: missing markets in environmental assets, government policy, failure to invest in human capital, and so on. But the problems are more serious when there is little or no conceptual framework at all, and that is the case with so much of the literature on "sustainability indicators". Simply put, they have overlooked the question of what these are meant to be indicators *of*. Hence we have spelled out one approach to defining sustainable development, and from the definitions the indicators follow fairly automatically. We have emphasised two areas where some success has been achieved: an economic definition based on the concept of "genuine savings" and perhaps overall wealth, and an ecological approach based on concepts of resilience and diversity. Both need further work, especially when we consider how they apply to individual sectors of the economy such as the agricultural sector.

Focusing on agriculture, we have asked how big a sector we should have, what it should produce and how it should produce it. The answers to all three questions involve us in exercises to identify the non-market attributes of agriculture. *First* we need to know what are its external benefits and costs — how much are we prepared to pay to conserve the countryside as we want it, and what is the trade-off between that and the benefits that agriculture provide both in terms of food, avoided environmental destruction, and actual amenity. *Second*, we need to understand the relationship between government policy and agriculture — to what extent is both the size and composition of the agricultural sector distorted by what we know are extensive and pervasive subsidies? *Third*, we need to understand the underlying cause and effect relationships, not just between, say, nitrogen run-off and eutrophication of water, but why the nitrogen run-off occurs in the first place. We have to explain why markets are good and bad: good because they give us what we want, and bad because they fail to identify and manage the non-market effects on the environment.

The agenda for the indicator business that emerges from this kind of analysis has a lot in common with the existing indicator business, but it also poses a rather fundamental challenge to improve that business as well.

BIBLIOGRAPHY

ANDERSON, J. and P. HAZELL (1989), *Variability in Grain Yields: Implications for Agricultural Research and Policy in Developing Countries*, Johns Hopkins University Press, Baltimore, United States.

ATKINSON, G. (1995), *Measuring Sustainable Economic Welfare: a Critique of the UK ISEW*, Centre for Social and Economic Research on the Global Environment, University College, London, Working Paper GEC 95-08.

ATKINSON, G., R. DUBOURG, K. HAMILTON, M. MUNASINGHE, D.W. PEARCE and C. YOUNG (1997), *Measuring Sustainable Development: Macroeconomics and the Environment*, Edward Elgar, Cheltenham, United Kingdom.

BATE, R. and R. DUBOURG (1995), *A Netback Analysis of Water Irrigation Demand in East Anglia*, Centre for Social and Economic Research on the Global Environment, University College, London, Working Paper WM 95-01.

CAMPBELL, L.H., M. AVERY, P. DONALD, A. EVANS, R. GREEN and J. WILSON (1997), *A Review of the Indirect Effects of Pesticides on Birds*, Joint Nature Conservation Committee, Peterborough, United Kingdom.

COSTANZA, R. *et al.,* "The Value of the World's Ecosystem Services and Natural Capital", *Nature*, Vol. 387, 15 May, pp. 253-260.

DALY, H.E. and J. COBB Jr. (1989), *For the Common Good*, Beacon Press, Boston, United States.

DEPARTMENT OF ENVIRONMENT, UNITED KINGDOM (1996), *Indicators of Sustainable Development for the United Kingdom*, HMSO, London.

EHRLICH, P. and A. EHRLICH (1997), "The Value of Biodiversity" in P. Dasgupta, K.-G. Mäler and A. Vercelli (eds), *The Economics of Transnational Commons*, Oxford University Press, Oxford, United Kingdom, pp. 97-117.

FOSTER, V., S. MOURATO, E. OZDEMIROGLU, D.W. PEARCE and S. DOBSON (1999), *The Social Value of the Charitable Sector*, Centre for Social and Economic Research on the Global Environment, University College, London.

HABERL, H. (1997), "Human Appropriation of Net Primary Production as an Environmental Indicator: Implications for Sustainable Development", *Ambio*, Vol. 26, No. 3, pp. 143-146.

HAMILTON, K. and E. LUTZ (1996), *Green National Accounts: Policy Uses and Empirical Experience*, Environment Department Papers, No. 39, World Bank, Washington, D.C.

HAMILTON, K., D.W. PEARCE, G. ATKINSON, A. GOMEZ-LOBO and C. YOUNG (1994), *The Policy Implications of Natural Resource and Environmental Accounting*, Centre for Social and Economic Research on the Global Environment, University College, London, Working Paper GEC, pp. 94-118.

HAMILTON, K., G. ATKINSON and D.W. PEARCE (1996), "Measuring Sustainable Development: Progress on Indicators", *Environment and Development Economics*, Vol. 1, No. 1, pp. 85-102.

HOLLING, C.S. (1973), "Resilience and Stability of Ecological Systems", *Review of Ecology and Systematics*, Vol. 4, pp. 1-24.

HAZELL, P. (1984), "Sources of Increased Instability in Indian and US Cereal Production", *American Journal of Agricultural Economics*, Vol. 6, No. 3, pp. 302-311.

JACKSON, T. and N. MARKS (1994), *Measuring Sustainable Economic Welfare: a Pilot Index 1950-1990*, Stockholm Environment Institute, Stockholm.

KUNTE, A., K. HAMILTON, J. DIXON and M. CLEMENS (1998), *Estimating National Wealth: Methodology and Results*, Environment Department Papers, No. 57, World Bank, Washington, D.C.

PEARCE, D.W., A. MARKANDYA and E. BARBIER (1989), *Blueprint for a Green Economy*, Earthscan, London.

PEARCE, D.W. (1993), *Economic Values and the Natural World*, Earthscan, London.

PEARCE, D.W. (1998), "Auditing the earth", *Environment*, Vol. 40, No. 2, March, pp. 23-28.

PEARCE, D.W. and G. ATKINSON (1993), "Capital Theory and the Measurement of Sustainable Development", *Ecological Economics*, Vol. 8, No. 2, pp. 103-108.

PEARCE, D.W., G. ATKINSON and K. HAMILTON (1998), "The Measurement of Sustainable Development", in J.C.M van den Bergh and M. Hofkes (eds), *Theory and Implementation of Economic Models for Sustainable Development*, Kluwer, Dordrecht, The Netherlands, pp. 175-193.

PEARCE, D.W. and R. TINCH (1998), "The True Price of Pesticides", in W. Vorley and D. Keeney (eds), *Bugs in the System: Redesigning the Pesticide Industry for Sustainable Agriculture*, Earthscan, London.

PERRINGS, C. and M. COMMON (1997), "Towards an Ecological Economics of Sustainability" in C. Perrings, *Economics of Ecological Resources: Selected Essays*, Edward Elgar, Cheltenham, United Kingdom, pp. 64-90.

PIMENTEL, D., H. ACGUAY, M. BILTONEN, P.M. SILVA, J. NELSON, V. LIPNER, S. GIORDANO, A. HAROWITZ and M. D'AMORE (1992), "Environmental and Economic Costs of Pesticide Use", *Bioscience*, Vol. 42, No. 10, pp. 750-760.

SCHELLING, T. (1998), *Costs and Benefits of Greenhouse Gas Reduction*, American Enterprise Institute for Public Policy Research, Washington, D.C.

SMITH, V.K. (1992), "Environmental Costing for Agriculture: Will it be Standard Fare in the Farm Bill of 2000?" *American Journal of Agricultural Economics,* Vol. 74, No. 5, December, pp. 1 076-88.

STEINER, R., L. McLAUGHLIN, P. FAETH and R. JANKE (1995), "Incorporating Externality Costs into Productivity Measures: a Case Study using US Agriculture" in V. Barbett and R. Payne (eds), *Agricultural Sustainability: Environmental and Statistical Considerations*, Wiley, New York, United States, pp. 209-230.

STIGLITZ, J. (1979), "A Neoclassical Analysis of the Economics of Natural Resources", in V.K. Smith (ed.), *Scarcity and Growth Revisited*, Johns Hopkins University Press, Baltimore, United States.

SWISS FEDERAL OFFICE OF AGRICULTURE (1998), *Summary Report on Multifunctional and Sustainable Agriculture (with reference to the next WTO round)*, Tänikon Seminar, 29-30 April, Berne, Switzerland.

VISCUSI, W.K. (1994), "Risk-risk Analysis", *Journal of Risk and Uncertainty*, Vol. 8, pp. 5-17.

VITOUSEK, P., P. EHRLICH, A. EHRLICH and P. MATSON (1986), "Human Appropriation of the Products of Photosynthesis", *Bioscience*, Vol. 36, pp. 369-373.

VITOUSEK, P., H. MOONEY, J. LUBCHENKO and J. MELILLO (1997), "Human Domination of Earth's Ecosystems", *Science*, Vol. 277, 25 July, pp. 494-499.

WORLD BANK (1994), *Population and Development*, Washington, D.C.

WORLD BANK (1997), *Expanding the Measure of Wealth: Indicators of Environmentally Sustainable Development*, Washington, D.C.

WORLD COMMISSION ON ENVIRONMENT AND DEVELOPMENT, [the Brundtland Commission] (1987), *Our Common Future*, Oxford University Press, Oxford, United Kingdom.

XIE, J. (1996), *Water Subsidies, Water Use and the Environment*, World Bank, Environment Department, Washington, D.C., mimeo.

PART II:

SUMMARY OF THE WORKSHOP DISCUSSION

AND RECOMMENDATIONS

OVERALL SUMMARY OF THE WORKSHOP DISCUSSION AND RECOMMENDATIONS

by
Chris J. Doyle
Scottish Agricultural College, Auchincruive, United Kingdom

1.   Background to the Workshop

OECD countries have attached high priority to developing a set of agri-environmental indicators (AEIs). This is a consequence of recent OECD Ministerial meetings at which the importance of integrating environmental concerns into agricultural policy reform has been emphasised by the *OECD Secretariat* paper in this report. To measure the progress in the agricultural sector towards sustainable management of natural resources and improved environmental performance, it is clearly necessary to develop a comprehensive set of robust indicators. To this end OECD has set itself three key objectives for developing AEIs, in order to meet the needs of policy makers as well as other groups with a stake in agriculture and the environment.[1] These are to:

- *provide information* to policy makers and the wider public on the current state and changes in the conditions of the environment in agriculture;

- *assist policy makers* to better understand the linkages between the causes and effects of the impact of agriculture and agricultural policy on the environment, and help to guide their responses to changes in environmental conditions;

- *contribute to monitoring and evaluation* of the effectiveness of policies in promoting sustainable agriculture.

A key challenge is that environmental indicators aim to capture the relationship between biophysical processes and human activities, unlike economic and social indicators which are primarily concerned with the measurement of human activities. AEIs additionally need to capture both spatial and temporal dimensions of environmental change. Furthermore, implicit in the concept of sustainable development are the linkages between social, economic and environmental dimensions, which AEIs need to capture. Lastly, as underlined by *David Pearce* in this report, if the trade-offs between economic, environmental and social phenomena are to be addressed, then it is necessary to develop AEIs that are expressed in a common monetary unit.

---

1.   See OECD (1997), *Environmental Indicators for Agriculture*, Paris.

Work on developing AEIs essentially involves four steps. These cover the:

- development of a conceptual and analytical understanding of the various agricultural and environmental processes;

- identification of appropriate indicators and methods of measurement;

- collection of data and calculation of the indices; and

- integration of indicators into policy analysis.

So far the OECD has developed a framework, the so-called *Driving Force-State-Response (DSR) Framework,* which seeks to address the linkages between causes, effects and actions. Alongside this it has identified thirteen AEIs. Accordingly, a key objective of the Workshop was to develop a set of indicators for those areas which are least advanced in terms of their conceptual and analytical basis and methods of measurement, namely:[2]

• water quality	• wildlife habitats
• water use	• landscape
• soil quality	• farm management
• land conservation	• farm financial resources
• biodiversity	• socio-cultural issues (rural viability).

This summary attempts to draw together the main threads that emerged from the papers presented at the Workshop and the ensuing discussions. While the primary focus was on selecting appropriate indicators (see Section 2), there was also considerable discussion on the application of the indicators for policy purposes (see Section 3). Accordingly, both these aspects are considered in this summary, together with a résumé of the discussion in the Workshop on proposing ways to develop and use AEIs in the future in the OECD (see Section 4).

---

2.   It should be noted that for *nutrient use, pesticide use and risk*, and *greenhouse gases*, OECD work is already underway, see Annex, but these areas were not examined at the Workshop.

2. Selection of indicators

2.1 Attributes of effective indicators

The work in the OECD has already established that, to be effective, AEIs must meet the following selection criteria as outlined in the *OECD Secretariat* paper:

- policy relevance — policy issue, rather than data driven;

- analytical soundness — based on sound science;

- easily interpreted — essential information communicated to users;

- measurable — data which can be realistically collected and measured, taking into account spatial and temporal considerations.

While there was a high degree of consensus at the Workshop regarding these selection criteria, questions were raised regarding whether they were sufficiently specific or comprehensive. Thus, a number of participants and speakers, including *David Pearce* and *Elliot Morley*, stressed that, while it was necessary for indicators to be policy-relevant, the policy context itself had to be spelt out.

A number of participants and speakers, including *Elliot Morley* and *Andrew Moxey*, proposed that two other criteria were equally important, namely public acceptability and ease of understanding. With regard to the former criterion, it was argued that if indicators did not gain general acceptance by the public, as well as by policy makers, it could be difficult to implement policies based on them. In respect of the latter criterion, it was remarked by several speakers that, unless indicators could be readily understood, it would be difficult to convince farmers that they should modify their management practices in response to them. However, some speakers were sceptical regarding the validity of this point, arguing that the behaviour of farmers and the farming sector is substantially influenced by the policy framework.

Several participants suggested that the policy context was "*sustainable development*" and that all the indicators selected should be scrutinised with respect to whether they contributed to this goal. However, some questioned whether the concept of "sustainable development" could be sufficiently articulated to be meaningful. Others also contended that, if too much emphasis was placed on "issue-driven" indicators, there was a real danger that factors such as feasibility of construction and measurability would be ignored. In particular, several participants argued that in the short run there was a need to be pragmatic and to accept that data availability was likely to be a key determinant of an initial set of indicators.

The discussion also highlighted problems with those indicators that involved comparing the "state" of a particular environmental pollutant with some reference or "threshold" value. It was recognised, as *David Pearce* underlined in his plenary addresses, that AEIs only become true indicators when they are compared with some meaningful benchmark. In addition, if AEIs are to have a real impact on policies, then they need to be forward-looking and to provide information on how far development is proceeding along a sustainable path. This in turn means that "thresholds" or "benchmarks" need to be defined for each indicator, without which they are not capable of meaningful interpretation.

Some participants, however, cautioned against presuming that such thresholds would be easy to define for many indicators. In particular, it was noted that a lack of understanding of ecosystem processes may inhibit the identification of scientifically valid "norms". Moreover, for several of the indicators considered, it was found to be impossible to establish a common threshold value, which was acceptable to all participants or applicable in all situations. However, if countries did use different reference values, then it would be difficult to make cross-country comparisons. It may therefore be important for OECD to urge that (for its purposes) a common, albeit arbitrary, benchmark level (e.g. a national standard for drinking water quality for each country) be used for each indicator, although this would require careful consideration of variations in regional and national base conditions.

2.2   *Identifying appropriate indicators*

Against this background three Groups were asked to identify a list of three to four key indicators for each of the agri-environment areas (the discussion and recommendations for each indicator area are discussed in greater detail in the summary reports below). Despite doubts expressed regarding the feasibility of identifying such a limited sub-set of indicators which had reasonably universal applicability, the three Groups were generally successful in this objective. However, the conclusions from the Groups differed significantly in the precision and comprehensiveness of the indicators they were able to recommend. This reflects to a large extent the fact that some agri-environmental areas have been the subject of more recent research (e.g. biodiversity, habitats and landscape) than others, for which knowledge has evolved over a longer period of study (e.g. soil and water issues).

It also emerged from the Group's discussion that a number of indicators were of a "contextual" nature. The key areas identified cover related agricultural trends in: land use and cover, farm population and structures (see summary list in the Annex). These indicators can provide both background information to trends in specific indicators and also help to emphasise the global approach needed to analyse agri-environmental issues.

Group 1: Water and Soil Quality, Water Use and Land Conservation

Relative to the others, this Group had a comparatively easy task, since there exists a long history of developing indicators such as for soil and water quality. The problem is not so much one of defining indictors as selecting from the considerable list of available ones. The finalised list of indicators for the soil and water areas, for both short- (immediate) and medium/long-term development, are given in Table 2.

Although the Group stated in its report that the DSR framework was used to inform the selection of key indicators, it was very difficult to identify indicators that embraced all three elements of the framework. In particular, it observed that it was especially difficult to identify indicators of "response", which were truly quantitative and unambiguous in their interpretation.

The Group also reported that it was not possible to construct indicators for which the interpretation would be similar in all situations, a theme to which the reports of all of the Groups returned. An example put forward of the problems of designing "universal" indicators was the index of "inherent soil quality" (see Table 2). Thus it was observed that knowing the capability of the soil at a particular site was of limited value in itself. The indicator would only acquire policy relevance if it is compared with something else, for instance the current land use at a site, and the stock and fluxes of soil organic carbon in agricultural soils. For example, any mismatch between the soil capability and existing land use would provide some valuable policy information.

In selecting indicators for these agri-environmental areas there was a strong tendency to focus on unambiguous biophysical measures at the expense of economic and social indicators. However, the problem with developing economic indicators, such as the "economic efficiency of water use", is that a number of issues need to be addressed before such indicators can have policy relevance. These include the effects of commodity price distortions caused by agricultural policies, and the valuation of externalities. In the case of social indicators, especially with reference to the impacts of soil and water management on agricultural employment and rural depopulation, there is an evident lack of consensus on such issues as: what is being measured; how to construct appropriate indicators; and their universal applicability.

Nevertheless, in certain OECD countries, water is a principal factor for assuring the viability of local communities, by avoiding depopulation, land abandonment and consequent land degradation.

Table 2. List of recommended indicators proposed for the soil and water areas

Water Quality	Water Use	Soil Quality	Land Conservation
• **Nitrate concentration** • **Phosphorus concentration** • **Risk of water contamination by nitrogen**	• **Water use intensity** • **Water stress**	• **Risk of water erosion** • **Risk of wind erosion**	• **Water buffering capacity**
• Risk of water contamination by pesticides	• Water use efficiency (technical and economic aspects) • Policy/management response to water stress	• Inherent soil quality	• Off-farm sediment flow

Indicators in **bold** are for short-term development and those unbolded are for medium- to long-term development. For detailed definitions of these indicators, see Annex.

Group 2: Biodiversity, Wildlife Habitat and Landscape

Defining indicators for these agri-environmental areas proved to be especially challenging, because of the complexity and lack of understanding of many of the ecosystem processes involved. Thus, there was general agreement within the Group that in view of the relatively recent advent of work on indicators in these areas, it would be premature and unproductive to seek to identify a clearly defined set of indicators at this time. Instead the discussions concentrated on identifying the conceptual and methodological issues relating to the establishment of appropriate indicator sets, and identifying an initial set of indicators for short-term development. These discussions revealed a number of common and cross-cutting issues, including the need to:

- develop indicators which recognise the key linkages between biodiversity, habitats and landscape;

- recognise the importance of a degree of flexibility for countries to choose indicators suited to their own agri-environmental circumstances, although this should not undermine the need for a consistent methodology to facilitate cross-country comparison;

- recognise the importance of the spatial resolution of developing these indicators;

- develop contextual indicators, such as those covering changes in land use and crop cover, to emphasise the global approach to analyse these areas;

- develop suitable baseline/threshold values to interpret changes in the indicators;

- be more specific about the policy questions at which indicators are aimed.

Against this background the Group confined itself to identifying the key components within each of the three indicator areas, and this led to the recommended list of "categories" for indicators, for both short- (immediate) and medium/long-term development, in Table 3.

Nevertheless, considerable work remains to be done at the specific indicator level in respect of data needs, methodologies and interpretation. In particular, there is a considerable degree of linkage between the areas, notably biodiversity and habitat. More specifically, in respect of biodiversity, the development of biodiversity quality indicators is not currently feasible, because of the lack of a standardised regime and the enormous complexity for choosing common species or taxonomic groups. However, there was a general view that surrogate measures, such as the state of the populations of certain indicative species (e.g. birds) and of the state of threatening processes (for example, with respect to exotic species or of pollution levels) could serve as useful proxies of biodiversity quality.

There was agreement that at this stage there is not so much a need for one single highly aggregated biodiversity indicator, such as the natural capital index, but rather demand for the underlying more detailed indicators, such as biodiversity quantity and quality. It was also stressed that baselines are indispensable for interpreting the state and trends in biodiversity, but since the choice of an appropriate baseline is complex, further work is required for their determination.

In respect of landscape, it was felt that it should be described in terms of a "set" of spatial units or landscape typologies, which in turn were a reflection of natural, land use and cover, cultural and management elements. Accordingly, the relevant indicators need to capture the land characteristics, cultural features and management functions that determine both the type and change in landscapes. But there is considerable work ahead before fully operational landscape indicators, suitable for policy purposes, will be available.

Table 3. List of recommended indicators proposed for the biodiversity, habitat and landscape areas

Biodiversity	Wildlife Habitat	Landscape
• **Genetic diversity of domesticated livestock and crops**	• **Intensively farmed agricultural habitats**	• **Land characteristics of agricultural landscape**
• **Wildlife species diversity related to agriculture**	• **Semi-natural agricultural habitats**	• **Cultural features of agricultural landscape**
	• **Uncultivated natural habitats**	• **Management functions of agricultural landscape**
• Change in numbers of endangered species related to agro-ecosystems	• Habitat heterogeneity (average size of habitats)	• Developing system of landscape typologies
• Impacts on biodiversity of different farm practices and systems	• Habitat variability (number of habitat types per monitoring area)	• Public surveys and monetary valuation of societal landscape preferences
• Effects on biodiversity caused by off-farm soil sediment flow	• Impacts on habitat of different farm practices and systems	

Indicators in **bold** are for short-term development and those unbolded are for medium- to long-term development. For detailed definitions of these indicators, see Annex.

Group 3: Farm Management, Farm Financial Resources and Socio-Cultural Issues

In order to provide a framework for developing indicators in respect of farm management, financial resources and socio-cultural issues, there was considerable debate on whether the aim was to develop "environmental" indicators for agriculture, or "sustainable agriculture" indicators. It was concluded that the latter properly encompassed the policy context. Four concepts for indicators covering this group were agreed on, namely:

- farm management capacity;

- on-farm management practices;

- farm financial resources;

- socio-cultural issues (rural viability).

The first two were seen as covering environmental sustainability of agriculture; the third and fourth are related to the economic and social dimensions, respectively. The indicators proposed by the Group, for both short- and medium/long-term development, are given in Table 4.

Concerning farm management indicators, the Group stressed that there was a need to distinguish between:

- the advisory and information inputs into farm decision making;

- the formulation of plans and strategies for the farm; and

- the environmental consequences for farming activities and practices.

Arguably there needs to be a hierarchy of indicators to reflect these distinctions. Furthermore, it was strongly argued that this hierarchy needs to be able to reflect not just levels of management, but also the spatial hierarchy in terms of farm, region and country. Equally significant was the observation that, due to the wide range of ecological, spatial and climatic conditions across countries, comparing indicator trends over time might be more meaningful and more practical than comparing absolute values. This, of course, is an observation that is also relevant for other indicators.

Farm management and farm financial indicators are important in that they can serve as proxies for state indicators. Because of data difficulties and costs of monitoring, it is frequently impractical to measure certain physical environmental states. Certainly, while basic quantitative data on land use, farm accounts, pesticide use and nutrient applications are available in many countries, the same is not true in respect of the environmental impacts associated with farming.

Table 4. List of recommended indicators proposed for the farm management, farm financial and socio-cultural (rural viability) areas

Farm Management		Farm Financial Resources	Socio-cultural (Rural Viability)
Farm management capacity	*On-farm management practices*		
• **Standards for environmental farm management practices**	• **Matrix of environmental farm management practices (with a substructure covering management of nutrients, pesticides, water, soil, etc.)**	• **Public and private agri-environmental expenditure**	• **Share of agricultural income in relation to total income of rural households**
• **Expenditure on agri-environmental research** • **Educational level of farmers**		• **Farm financial equilibrium**	• **Entry of new farmers into agriculture**
• Number of agricultural advisers trained in environmental management practices	• Implementation index to express the results of the matrix of environmental farm management practices	• Adjusting farm financial resources for changes in natural resource depletion and pollution	• Social capital in agricultural and rural communities

Indicators in **bold** are for short-term development and those unbolded are for medium- to long-term development. For detailed definitions of these indicators, see Annex.

Management indicators are also important because they are the "driving forces", which relate directly to agriculture. In contrast, changes in "state" indicators, such as water quality, reflect both agricultural and non-agricultural impacts. Additionally, farm management indicators can provide an early indication of changes underway, frequently well before the actual consequences are apparent from agri-environmental state indicators. Management indicators are also useful in their own right for assessing the relationship between farm practices and the state of the environment. Monitoring trends in indicators of management practices, alongside appropriate state indicators, will allow policy makers to better evaluate the success of policies aimed at the environment.

Considerable work is still needed to further refine many of the proposed indicators in this Group, before they can be fully operational. There was also some concern expressed by participants about both the coverage and transparency of the proposed indicators. Thus, several participants noted the need for indicators relating to agricultural employment, specifically in respect of measuring socio-cultural issues (rural viability). Concern was expressed about the transparency of some of the indicators. Perhaps more than for other AEIs, there is a need for farmers, and not just policy makers, to be able to understand, relate to, and interpret these indicators. In particular, if changes in farm management practices are to be encouraged as a result of the indicators, it will be important that the validity of the indicators and their interpretation are widely accepted by the general public.

2.3   How much progress?

At the outset of the Workshop, the OECD had stated that the key objective was to identify a small number of AEIs, which were both widely applicable and relevant to agricultural policy making. Although a number of participants expressed important reservations about whether the task was practical, given the diversity of environmental problems facing OECD countries, it has to be said that the discussions on specific indicators were broadly successful. Notwithstanding that the progress was not uniform across the different indicator areas, the key issues were identified and workable frameworks were outlined on which to build the development of AEIs. In particular, a significant number of new proposals for indicators were advanced at the Workshop.

Although, in terms of its objectives the Workshop could be judged to have made progress, a number of significant issues remain unresolved and will need to be returned to at a later stage of the work. Among these was the issue of linkages between the different agri-environment areas, and it will be necessary to assess fully the degree of overlap among the recommended indicators, and to explore the linkages between the indicators proposed in one area and those in other areas. If collectively the suite of AEIs put forward by OECD for its thirteen areas are to provide comprehensive coverage of agri-environmental interactions and form a coherent framework, then work needs to be completed on the cross-linkages relating specific indicators to the different indicator areas.

At the same time, in terms of the DSR framework advanced by OECD, it is arguable that too many of the recommended indicators are concerned with measuring the state of the environment and, within this framework, biophysical and technical phenomena. In particular, it was stressed that for AEIs to have a real impact on policies they need to be forward-looking and to provide information on how far development is proceeding along a sustainable path. This implies that indicators should be giving an indication of the environmental "risk" involved. In this way, policy makers will have an opportunity to prevent environmental degradation rather than simply implement policies to repair the damage.

Greater attention also needs to be addressed to developing indicators which explicitly integrate biophysical with human phenomena. Specifically, several participants called for indicators which measured the degree of interaction between government agencies, the public and farmers in respect of environmental actions and policies. Of course, indicators that reveal these interactions are conceptually difficult to construct and potentially ambiguous.

Another issue, which requires more attention, concerns the level and method of aggregating indicators. Specifically, the objective of the OECD is to develop indicators which can be interpreted at the national level. This often involves both spatial aggregation and the integration of indices measured in different units. Both *David Pearce* and *Andrew Moxey,* emphasised that, in spite of widespread concerns, there were no real alternatives to converting everything to a single monetary measure. In particular, it is very difficult if not impossible to undertake an analysis of the trade-offs between the environment and economic activity unless environmental costs and benefits are expressed in monetary equivalents. However, this did not prevent several participants questioning whether such an approach could capture all the dimensions of sustainability and attention was drawn to the fact that some countries had used energy units as the common currency.

In summary, the Workshop was universally accepted to have advanced thinking and to have been successful on several fronts. The immediate future objectives must be to:

- refine further some of the proposed indicators;

- identify potential gaps in terms of the ideal coverage of AEIs and in terms of the DSR framework;

- improve the overall coherence of the indicator sets by mapping out the cross-linkages and removing any overlaps and duplication between areas.

3.   **Application of agri-environmental indicators for policy purposes**

The development of AEIs is of course only a first step towards their use for policy purposes. In this context the papers by *David Baldock* and *Paul Thomassin* explored the issues of the application of AEIs in policy making and analysis, and it is convenient to summarise these presentations and the ensuing debate in terms of three issues:

- the role of AEIs for policy purposes;

- the potential problems relating to the development and application of indicators;

- the integration of indicators into socio-economic impact analysis as predictive tools.

3.1   The role of agri-environmental indicators for policy purposes

The use of indicators as an aid to policy decision-making in the agri-environmental context is a relatively recent phenomenon and still a developing field, as *David Baldock* admitted, a set of standard international indicators has yet to emerge. However, indicators are perceived to have considerable potential as policy tools.

Most policy makers concerned with agri-environmental issues at the national level, are confronted with fragmented information and it is accordingly difficult to harness the information in a way that effectively contributes to policy decision making. Many governments are beginning to invest in indicators as tools to aid policy making in a systematic way. Currently wide use is made of AEIs in:

- state of environment reports;

- sustainable development reports;

- responses to international agreements;

- monitoring progress in meeting agri-environment policy objectives;

- evaluation of the impacts of particular agricultural and agri-environment policies;

- ranking of bids for agri-environmental schemes; and

- longer-term predictions of agri-environmental policies.

However, while indicators are being introduced into the policy-making process, they are being included in an ad hoc way in response to short-term policy pressures. Many of these pressures arise from new legislation and initiatives, which have introduced requirements to undertake evaluations and meet specific targets in respect of domestic agri-environmental schemes and international environmental agreements. Even where a more coherent and long-term approach is adopted, involving the DSR approach advocated by the OECD, the policies themselves are not always clear or specific. Part of the problem is that agri-environmental policies may be seen both as "driving forces" and as "responses" to specific environmental issues, and may also contain objectives which are not purely environmental, such as supply control. As such this underlines the fact that the DSR framework needs to be used flexibly and not treated as a rigid structure constraining the choice of indicators.

3.2   *Potential problems and dangers of developing and applying indicators*

As underlined by several participants, for most policy makers indicators need to be simple, clear and meaningful. In addition, they have to be reliable and able to be compiled without heavy expenditure on collection of new datasets. They also need to be credible to the wider public. However, within the OECD Member countries, there is a great range of agricultural systems and practices. The relationship between these practices and the many ecosystems in which they are applied is diverse and complex. The intricacy of the linkages between agricultural activity, the environment and policy also introduces substantial technical and intellectual challenges, as well as certain hazards. In particular, some of the key technical and conceptual issues relating to the selection and design of AEIs centre on data issues, indicator construction and maintenance of the indicator data sets. Each of these are considered separately.

Data issues

The availability of relevant data on agricultural systems and on the environment varies across OECD countries. Moreover, many indicators have been constructed from national data, which are not intended to reveal the specific contribution of agriculture to environmental change. Thus, fresh research may be required to develop more appropriate indicators for agri-environmental policy making.

60

Shortages of baseline data have inhibited, and will continue to inhibit, work on developing indicators. Notably, precise information about farm management and changing practices is in limited supply in many countries. This makes it difficult to develop sophisticated indicators, which capture critical aspects of the impact of management on the environment. However, some participants sounded a cautionary note, observing that countries need to review the wealth of data already collected, before embarking on collection of new data sets for agri-environmental purposes. Finally, it has to be recognised that there has also been a temptation to select certain indicators, not because of their policy relevance, but because the data are available. However, this may in part reflect countries concerns about enlarging data demands when budgetary constraints for many countries act against collection of new data sets.

Indicator construction

In respect of indicator construction, a number of dangers were identified:

- It can be difficult to develop meaningful quantitative indicators for certain agri-environmental areas at a national level because of the diversity of natural and cultural conditions within a country. National level quantification is often easier with respect to basic resources, such as soil, water and air. However, for other environmental areas, such as biodiversity, habitats and landscape indicators are of little value without reference to the spatial context.

- Work on sustainable agriculture brings together social, economic and environmental indicators and the temptation is to try and integrate all three elements by developing "composite" indicators. However, this often produces highly aggregated and meaningless indicators.

- For policy makers, it is very important to be clear about the precise linkage of any indicator to agricultural practices. A number of indicators have been developed which are not very sector specific, in the sense that agriculture's contribution to the harm and/or benefit of the environment cannot be separated from other economic sectors. This makes it difficult to interpret the agricultural policy implications of these indicators.

- As governments widen their preoccupation from the environmental impacts of farming to issues of animal welfare and food safety, it will be important to recognise that AEIs are only one element in a wider spectrum of information for policy makers and other stakeholders.

- A number of international agencies are developing AEIs and there is thus a real need to exchange information on developments in this area to avoid duplication.

- It is clear that OECD needs to accept that it cannot progress with the development of indicators at a faster rate than individual OECD countries and that this will mean that not all of the thirteen proposed indicator areas can advance at the same speed.

Maintenance of indicators

One of the major focal points of the discussion was on the development of actual indicators, while less attention was given to the maintenance of the indicators and appropriate data sets, once constructed. It was stressed that there is a need to recognise the fact that certain indicators could become obsolete as policy circumstances change and therefore the policy relevance of all indicators should be periodically reassessed. The implication of this is that the OECD will need to institute procedures for regularly reviewing indicators, and removing and replacing indicators as necessary to better reflect the changing policy environment.

3.3   *The integration of indicators into socio-economic impact analysis as predictive tools*

Generating AEIs is clearly not an end in itself. In particular, indicators are by their nature backward-looking, in that they are based on historical data. But as some participants pointed out, policies are essentially looking to the future. Thus, indicators must be able to generate information on social and economic impacts. For this reason *Paul Thomassin* strongly argued that AEIs had to be integrated into public policy models, if they were really to contribute to decision making.

Certainly, AEIs have been used in a variety of modelling exercises to increase the quantity and quality of information available for evaluating policy impacts. Moreover, as several participants stressed, policy simulation models offer a way of getting the maximum benefit from existing data, without the need to collect costly new data. Nevertheless, there is a need to recognise that models are only as good as the data they use, their conceptual underpinning and their scientific soundness, and it is essential that they are assessed for their analytical reliability before they are applied.

However, if indicators are to be integrated into policy models and used in this way several issues need to be considered:

- It will be necessary to test changes in the indicators against the parameters used in specific models to verify their suitability for modelling purposes. In this respect some "state" indicators may be unsuitable.

- Clear causal links between the policy actions and resource impacts need to be identified, so that the links between interpreting the indicators, developing appropriate policy responses and predicting the environmental and economic impacts can be clearly quantified.

- Given the spatial heterogeneity of many environmental resources, any models developed must be sufficiently detailed in order to identify the spatial distribution of the impacts.

- In predicting future impacts of policies, it is vital to recognise that the behavioural responses of farmers will be conditioned not only by agricultural policies themselves, but also, in particular, other economic policies and by the range of technological options available. This can add greatly to the uncertainties of prediction.

3.4   Concluding remarks on using indicators for policy purposes

Although several participants expressed reservations about the use of models to predict policy impacts, there was considerable consensus that the issue was not whether models should be used, but how to make them more effective and credible. To this end it was stressed that the key to constructing successful models was an iterative dialogue between scientists, economists, modellers and policy makers. Furthermore, model development needs to proceed in tandem with data development, so that appropriate proxy variables used in the modelling process can be identified. However, dialogue is only one element in developing successful agri-environmental policy models, there are also certain conceptual and methodological issues that need to be resolved, including:

- Integration of economic and biophysical data, so that the links between economic and environmental activities can be mapped.

- Enhanced understanding of the behavioural response of decision makers, such as farmers, to environmental signals provided by AEIs.

- Endogenous treatment of both the economic and environmental aspects of policy.

- Integration and interfacing of models developed at different system levels (e.g. farm, region and country).

In the longer term, models will need to be capable of adjusting to the information requirements of decision makers. This will necessitate a much greater emphasis on configuring the model output. Mechanisms will have to be developed, therefore, which link the information output from the integrated models to the responses of decision makers. Only in this way will decision makers actively address potential environmental problems before they occur. As underlined earlier, prevention rather than restoration of environmental degradation must be the strategic objective.

4.   The way forward

The closing statements by *Gérard Viatte* and *Dudley Coates* demonstrated that, while the Workshop did reveal differences in the philosophies of the various participants, it also generated a very useful and potentially productive discussion on the issues related to the identification and application of AEIs. Advances had been made in both the identification and construction of policy-relevant indicators, and the recommendations from the Workshop will provide a framework for the OECD to progress these issues further. Perhaps above all it uniquely brought together policy makers, researchers, international governmental organisations and non-governmental organisations, providing an insight into the concerns and priorities of both the constructors and users of indicators.

More specifically, the main achievements of the Workshop may be summarised as follows:

- It identified a set of indicators, for both short-term (immediate) and medium/long-term development, which command broad consensus in terms of feasibility and policy relevance. However, it is necessary to be pragmatic and to accept that the process will be one of evolution and refinement.

- While some of the indicators proposed are of the nature of "contextual data sets", such as land use and land cover, these will be valuable in interpreting trends in agricultural sustainability.

- It revealed the advantage of drawing on on-going work on indicators being developed in OECD Member countries as a basis to establish a common indicator methodology that is applicable to all OECD countries.

- In general, the indicators identified have high policy relevance and although some present immediate difficulties in terms of either data or conceptualisation, clear priorities have been established for future research across the different indicator areas.

- There was recognition that there will need to be some flexibility as regards the development and application of AEIs. Thus, a number of the indicators proposed are of a composite nature, having a set of varying elements, such as the inherent soil quality indicator covering various soil degradation processes, some of which are more relevant to certain countries than others.

- There was an acceptance that any indicators selected will need to be transparent, so that all "stakeholders" can understand them and accept the policy implications based on them.

It was also clear, however, that to make further progress it will be necessary to address some unresolved issues. In particular, the OECD will need to look closely at "cross-cutting" issues by examining the connections between the indicators proposed for the different areas. The further work will benefit from drawing on indicators being formulated both within OECD (such as on rural development) and by other international agencies. Also, there is arguably a need to revisit the rationale behind the thirteen indicator areas selected by the OECD, especially where some participants felt that some other areas might be included, such as the use of energy by agriculture.

Finally, several participants argued that the work should be broadened to encompass indicators of sustainable agriculture by specifically including the economic and social dimensions of sustainable agriculture. To some extent, however, this is already included in the OECD set of AEIs through the farm financial resources and socio-cultural indicators. Indeed, some participants suggested that the AEIs might be more accurately renamed as indicators of sustainable agriculture.

**SUMMARY REPORTS[3] AND RECOMMENDATIONS
COVERING SPECIFIC AGRI-ENVIRONMENTAL INDICATORS**

1. Common and cross-cutting issues for agri-environmental indicators concerning: water quality, water use, soil quality and land conservation

Basic quantitative data, such as annual precipitation, water quality, land use and inherent soil quality are available in many countries and regions, although they are not always comparable due to different definitions and/or methods of measurement. Coefficients to relate these data to the environmental impacts of agriculture are often missing or at an early stage of development.

There was a strong tendency to focus on unambiguous biophysical measures at this stage of the work, rather than policy and management responses, or economic and social indicators. Since they were still at a quantitative stage of development, *policy* and *management response indicators*, or in other words, "institutional indicators", were regarded as too lacking in rigour for the purposes of implementation, although it was accepted that it was desirable that such indicators should be developed with a more qualitative aspect. Thus, while this type of indicator will be studied further, the focus will be initially on its applicability with respect to water use.

Economic indicators could be developed provided that the effects of commodity price distortions, occurring as a result of agricultural policies, and the valuation of externalities, are addressed. There are relatively few attempts to value the economic costs of the impairment of water quality resulting from agricultural activities, or the economic benefits from particular management practices on land and water conservation.

Social indicators could be developed in the context of sustainable development, with particular reference to the impacts of soil and water management on employment and on the rural population

Regarding the coverage of indicator areas, it was accepted that soil quality and water use indicators should focus on the on-farm impacts of agricultural activity, while indicators for land conservation should focus on measuring the off-farm environmental consequences.

---

3.   The summary reports for each agri-environmental indicator area are based on reports provided by authors from OECD countries, see author list in Part IV, Annex, of this report — "Workshop Agenda and Main Contributors".

1.1   *Water Quality*

Policy issues

For most OECD countries, the principal sources of water pollution from agriculture include nutrients, pesticides and soil sediments, although problems of acidification, salinisation, and organic, biological and heavy metal contaminants associated with agriculture are important for some countries and in certain sub-national regions. An excessive level of nitrates in water is a human health concern since it impairs drinking water quality, while excessive concentrations of both nitrate and phosphorus cause ecological problems including eutrophication. Contamination of water by certain pesticides can be harmful to human health and aquatic life. The issue of soil sediment accumulation in off-farm surface waters is treated under Land Conservation, below, although it is not associated with human health problems.

While agriculture is not the only sector which burdens the quality of water with emissions of nutrients and pesticides, for most OECD countries it is the major cause of these problems. Although considerable efforts have been undertaken to diminish emissions of water pollutants from agricultural activity, the pollution of many surface waters and groundwater is deemed to be too high, especially in regions with intensive agricultural activities.

The key policy objective for improving the quality of water in agriculture is to contribute to the improvement of the suitability of water to maintain various human and ecological uses.

Main points discussed

The key areas of concern regarding the state of water quality identified relate to:

1. nitrate pollution in both surface and groundwater;

2. phosphorus levels in surface water;

3. the level of contamination with pesticides.

The initial proposals in the background paper involved the adoption of three indicators relating to the **state** of water quality in respect of these substances, in terms of the ratio of measured pollutant concentration in water to some defined reference level. However, it rapidly became evident that it would not be possible to agree a common reference level acceptable to all countries, as the "norms" would vary by country and also by the purpose of water use (e.g. human consumption, irrigation, etc.). Thus it was proposed that the indicators should be expressed as "the proportion of sampled water above an arbitrary but scientifically sound national standard rather than a single international standard". This would permit comparisons not only across time, but also across countries.

In developing such *state* indicators, one of the issues was whether the data should be collected on a nation-wide basis or monitoring should be confined to agricultural "vulnerable" areas. These areas could be defined using criteria such as livestock densities, proportion of crops requiring high nutrient applications, presence of sandy soils and steep slopes, and the intensity of irrigation.

It was suggested that for policy purposes indicators which identified those agricultural areas which were vulnerable, or at risk, would be very valuable. In this context it was accepted that developing *risk* indicators was useful to establish a clearer link between agriculture production, farm management practices and water pollution. Furthermore, such risk indicators would allow vulnerable areas to be identified.

However, the methodology underlying the construction of such *risk* indicators varies across countries and an agreement on a common methodology is yet to be reached. It was noted that those countries which were covered by the Oslo and Paris Conventions for the Prevention of Marine Pollution (OSPARCOM), are developing a nitrogen leaching risk model, and some of the principles embodied in this might be used to refine a model proposed by Canada, which only employs a nitrogen surplus risk calculation. As a first step, it was proposed that a nitrogen risk surplus model be applied to other countries as a pilot study, relying on national data on soil surface nitrogen balances already available at OECD.

The consensus reached was that both the *risk* and the *state* approaches deserve consideration as field data on the state of water quality will help to confirm risk assumptions. At the same time the risk of water contamination will facilitate identification of agricultural vulnerable areas and interpretation of the main causes of contamination.

Recommendations

Indicators for the short term

Nitrate Concentration in Water in Agricultural Vulnerable Areas

- *Definition*: The proportion of ground and surface water in agricultural vulnerable areas, above a reference level of nitrate concentration (NO_3 mg/l).

- *Method of calculation*: Sample concentrations of nitrate (mg/l) for groundwater and flow weighted mean concentrations (or mean concentrations per year) of nitrate (mg/l) for surface waters, in areas vulnerable to contamination from agriculture, directly measured by national authorities.

- *Interpretation*: Annual trends can be distorted by climatic hazards such as flood or drought. It would be useful to evaluate the extent to which multi-annual trends are linked to policy measures to reduce water contamination by nitrogen, including measures specific to vulnerable areas.

- *Further refinement*: The reference level (scientifically sound national standard, or a single international standard) and some minimum level of sample frequency, need to be defined. It is also necessary to establish the policy-relevant criteria to determine what constitutes a vulnerable agricultural area.

Phosphorus Concentration in Water in Agricultural Vulnerable Areas

- *Definition*: The proportion of surface water bodies in agricultural vulnerable areas, above a reference level of phosphorus concentration (P_{total} mg/l).

- *Method of calculation*: Flow weighted mean concentrations (or mean concentrations per year) of phosphorus (mg/l) for surface waters in areas vulnerable to contamination from agriculture directly measured by national authorities.

- *Interpretation*: Annual trends can be distorted by climatic hazards such as flood or drought. It would be useful to evaluate the extent to which multi-annual trends are linked to policy measures to reduce water contamination by phosphorus, including measures specific to vulnerable areas.

- *Further refinement*: The reference level (scientifically sound national standard, or a single international standard) and some minimum level of sample frequency need to be defined. It is also necessary to establish the policy-relevant criteria to determine what constitutes a vulnerable agricultural area.

Risk of Water Contamination by Nitrogen

- *Definition:* The area of agricultural land potentially at risk to water contamination by nitrogen.

- *Method of calculation:* The potential concentration of nitrogen in the water flowing from a given agricultural area, both percolating water and surface run-off, estimated as:

$$PNC \text{ (mg/l)} = PNP \text{ (mg/ha)} / EW \text{ (l/ha)}$$

where: PNC = potential nitrate concentration (mg/l)

PNP = potential nitrate present: (mg/ha) (for a given area, "excess nitrate" multiplied by the ratio of "excess water" to "soil water holding capacity plus excess water")

EW = excess water (l/ha) (precipitation less evapo-transpiration by crop type).

In order to identify the agricultural areas at risk, the potential concentration is to be estimated for each area with broadly uniform soil type and climatic conditions. However, as a first immediate and very preliminary step, the calculation of the potential nitrate concentration will draw on the OECD database on the soil surface nitrogen balance for agriculture at the national level. "Excess water" can be estimated from long-term (30-year) average precipitation and evapo-transpiration data (to be provided by national authorities or, when not available, drawing on the OECD Environment database). Data/coefficients on soil water holding capacity can be provided by national authorities, and possibly drawing on coefficients generated by the land conservation water buffering capacity indicator.

- *Interpretation:* Those areas with potential nitrogen concentrations above a reference level are regarded as vulnerable. Trends in areas at risk will mainly draw on changes in nitrogen surplus in each area, and to lesser extent changes in agricultural area.

- *Further refinement:* The methodology needs to be refined and assessed for its applicability at the regional and national level, and as a first step to undertake a national "pilot" test. It is intended that, in a second step, calculation of potential nitrate concentration will be refined by drawing on regional (sub-national) data (to be provided by national authorities).

Further development of indicators

Risk of Water Contamination by Pesticides

- *Definition:* The area of agricultural land potentially at risk to water contamination by pesticides

- *Method of calculation:* The potential concentration of pesticides in the water flowing from a given agricultural area, both percolating water and surface run-off, estimated as:

$$PCC_c \ (mg/l) \ = \ PCP \ (mg/ha) \ / \ EW \ (l/ha)$$

where: PCC   =   potential contaminant concentration (mg/l)

             c   =   index of the contaminant type (i.e. type of active ingredient)

        PCP   =   potential contaminant present: (mg/ha)

        EW   =   excess water (l/ha) (precipitation less evapo-transpiration by crop type).

- *Refinement*: Interpreting the risk of water contamination by pesticides is hampered by poor disaggregated pesticide use data, and a wide range in pesticide toxicity from drinking water for humans and for aquatic life. The indicator construction may benefit from on-going work in the OECD on agricultural pesticide risk indicators.

1.2   *Water Use*

Policy issues

Water underpins most aspects of human life. It is becoming increasingly clear that the availability of water is now a substantial limiting factor on the health and welfare of the global population. Irrigation has been used in many countries to extend the level of agricultural production where the natural rainfall pattern is at variance with crop needs. The extent of the pressures on total water resources and the consequent impacts on ecological processes vary from region to region reflecting, in many instances, population pressures, availability of water, and technological developments.

The key policy objective for the management of water in agriculture is to ensure that there are adequate water resources to meet the demand from agriculture and other users, with consideration of the social and environmental impacts of agricultural and other water uses, whilst improving the efficiency of water use by agriculture and sustaining aquatic ecosystems.

Main points discussed

The initial choice of indicators was determined by considering the key policy concerns in this area, which could be expressed in the form of three inter-related questions. These were:

1. the proportion of total available water diverted for all uses (and the contribution of agriculture to total water use);

2. the adequacy of provisions made for environmental demands on water;

3. the response of water resource managers to the challenge of providing adequate and efficient water supplies to agriculture taking into account environmental and economic considerations.

It was proposed that there should be three indicators, which roughly correspond to the DSR framework. The first was *water use intensity*, as measured by the proportion of water diverted for agricultural use to total available water resources. The second, termed *water stress*, measured the proportion of water resources subject to diversion for agricultural use for which there was no "Defined Minimum Reference Flow". The final indicator sought to capture the element of response, in terms of the presence or absence of a policy and/or management response to water stress.

Water use intensity needs to address measurement issues concerning how estimates of available water resources can be obtained. Generally, stream gauging, rather than estimations based on precipitation and evapo-transpiration calculations, was the preferred method. The concept of availability also presented methodological issues, regarding the handling of groundwater resources and fossil water. For water stress, the estimation of Defined Minimum Reference Flows also requires further thought.

The consensus was that the response indicator were regarded as too lacking in rigour for the purposes of implementation, although it was accepted that it was desirable that such indicators should be developed with a more quantitative aspect. This led to attention being focused on measures of *water use efficiency*, both technical and economic. In so far as the efficiency of the use of water for agricultural production can be used to indicate the extent to which pressures on water resources may be modified by technology and management, such measures provide an indication of possible future developments.

Some countries questioned the acceptability of aggregating all types of water (groundwater, surface water, fossil water) and contended that separate indicators, identifying the origin of water, were needed. However, it was deemed that this was mainly a national rather than an OECD concern.

One issue, that was only partially resolved, related to whether there was a need for socio-economic indicators in relation to water use. Such indicators, it was suggested, might capture the employment benefits of irrigated farming and the stabilisation of the rural population.

Recommendations

Indicators for the short term

Water Use Intensity

- *Definition*: The proportion of water resources subject to diversion for agricultural use.

- *Method of calculation*: The percentage of abstractions for all uses in total available water resources, with an indication of the share of irrigation water in total abstractions. The calculation requires (a) stream gauging data to be provided by national authorities or, when not available, mean annual precipitation less mean annual evapo-transpiration, plus (b) transborder water supply to estimate the annual freshwater availability (megalitres), and (c) the extent of freshwater abstractions (megalitres) for major uses, for example, irrigation, public water supply, industry and electricity.

- *Interpretation*: Interpretation should focus on pluri-annual trends in the share of agriculture in total abstractions. These may reflect changes in irrigated area and the composition of agricultural production although the share is also influenced by changes in water diverted by sectors other than agriculture. Annual trends are distorted by fluctuations in climatic conditions, and these trends should be interpreted in the context of trends in national water use intensity, for all uses.

- *Further refinement*: The indicator can be further refined to facilitate cross-regional/national comparisons where there are common land and climatic conditions or where there is dependence on a common water resource. The indicator would need further refinement to assess abstractions against a measure of the "divertible" or "renewable" water resource, and consider over-exploitation of groundwater resources.

Water Stress

- *Definition*: The proportion of rivers subject to diversion for irrigation without Defined Minimum Reference Flows.

- *Method of calculation*: The percentage of river lengths that do not have recommended minimum flow rate reference levels, based on information on regulatory measures (data on provisions for minimum flow rates in rivers to be provided by national authorities).

- *Interpretation*: This indicator assesses the risk of environmental damage through the absence of provisions for meeting environmental needs in those river system subject to diversion for agricultural use.

- *Further refinement*: The indicator can be further refined to assess the proportion of river lengths below the Defined Minimum Reference Flow, and also, more importantly, assess the adequacy of the Defined Minimum Reference Flows in meeting environmental needs.

Further development of indicators

Water Use Technical Efficiency

- *Definition*: For selected irrigated crops, the mass of agricultural produce (tonnes) per unit of the volume of irrigation water consumed, the latter being the volume of water, in megalitres, diverted or extracted for irrigation, less return flows.

- *Refinement*: Annual data on the volume of consumed water for selected irrigated crops, as well as yields of irrigated crops, need to be collected for a certain period of time. In order to remove the annual fluctuations caused by changes in climatic conditions, interpretation needs to focus on pluri-annual trends which may reflect changes in irrigation practices and trends in crop productivity. The indicator can be refined to assess the relative efficiency of different irrigation systems (flood, trench, sprinkler, dripper, etc.). It could also be further refined by evaluating the efficiency of storage, diversion and delivery systems (i.e. water lost in conveyance systems).

Water Use Economic Efficiency

- *Definition:* For all irrigated crops, the monetary value of agricultural production per unit irrigation volume consumed, the latter being the volume of water, in megalitres, diverted or extracted for irrigation, less return flows.

- *Refinement:* Annual data on the volume of water consumed for all irrigated crops, as well as the value of irrigated products, need to be collected for a certain period of time. In order to remove the annual fluctuations caused by changes in climatic conditions and commodity prices, interpretation needs to focus on pluri-annual trends which may reflect changes in the selection of crop type irrigated, irrigation practices and trends in crop productivity. The indicator needs to be refined to exclude the effects of price distortions resulting from agricultural policies.

Policy and Management Response to Water Stress

- *Definition*: The indicator would be an attempt to reveal the potential economic distortions in the use of water caused by under-pricing, free access or government intervention in the management of irrigation water, in particular, in countries or regions with a high intensity of water use.

- *Refinement*: It is recognised that in the short term measurements are difficult because of the need to establish the various components of comprehensive policy and management response indicators regarding water stress. Components might include cost recovery of water supply to agriculture, community involvement in water management, etc. There is also a need to identify the costs and benefits associated with the environmental impacts of water diverted for irrigation. Indicator construction may benefit from work underway by OECD on water pricing and subsidies.

1.3   *Soil Quality*

Policy issues

The limitations on the availability of soil resources to provide safe, nutritious food for an expanding world population is a critical issue when considering global food security. High quality soils are rare and at risk of degradation and loss through, for example, urbanisation. Although soil degradation is recognised as a widespread problem, for many countries the total area affected has yet to be determined. Soil losses due to environmental or climatic disasters as well as slow insidious changes in soil quality are serious concerns needing immediate attention.

The main potential adverse impacts from agricultural activities on soil quality include soil erosion, organic matter loss and the loss of soil biodiversity. Other impacts are important in some regions and for certain countries. These include soil contamination from the use of farm chemicals (including heavy metals), soil compaction (and structural decline), acidification, salinisation and waterlogging.

The key policy objective for the management of soil in agriculture is to optimise the appropriate functioning of soils as a limited resource for sustainable agricultural production, in ways that are environmentally safe, economically viable, and socially acceptable.

Main points discussed

Four key indicator groups were identified:

1. Inherent soil quality (linking soil capability to actual use, with low class values reflecting areas subject to degradation processes).

2. Risk of soil degradation by individual processes (water erosion, wind erosion, compaction, waterlogging, salinisation).

3. Soil biological qualities (soil organic matter and its decomposition, soil biodiversity).

4. Modelling soil behaviour in relation to stresses or responses (productivity loss, nutrient availability, acidification).

It was a consensus view that the key question was "what is the proportion or area of the land where current land use exceeds the assessed capability of the land?" In this context, it was accepted that an indicator of *inherent soil quality*, if it was combined with information on land use, was highly policy-relevant. Thus, there was a proposal to develop a map of inherent soil quality, possibly using one already developed by the United States as an example for the first stage of work. By comparing this "capability" map with one of land use, it would be possible to identify areas of mismatch, and to focus the attention of policy makers on areas which were at risk from soil degradation. This indicator should be regarded as a composite indicator, which covers various degradation processes relevant to each country, including the stock and fluxes of soil organic carbon in agricultural soils.

Among the various soil degradation processes, *soil erosion* was identified as a key policy issue, with wide policy relevance. Accordingly, producing indicators of the physical processes of quality degradation through water and wind erosion were considered to be a priority. It was felt that, for the time being, other impacts such as soil compaction, salinisation and waterlogging would be adequately covered in the inherent soil quality indicator, as these issues are relevant only to specific regions in certain countries.

There was some debate about whether there was a need to expand the scope of the soil quality indicator area by considering not only the biomass production function of agricultural soils, but also other "ecological" functions, such as the filtering, buffering and transformation of water. It was concluded that these concerns about off-farm environmental impacts should be dealt with separately in the Land Conservation indicator area (see below).

It was accepted that it is premature to measure soil biological qualities, although there is a need to start developing sound methodologies for this purpose. The modelling of soil behaviour also could be achieved, but there is a need for co-ordination across countries to ensure consistency in this respect.

Recommendations

Indicators for the short term

Risk of Soil Erosion by Water

- *Definition*: The agricultural area subject to water erosion (i.e. the area for which there is a risk of degradation by water erosion above a certain reference level).

- *Method of calculation*: Index of water erosion risk is calculated by the widely used Universal Soil Loss Equation (USLE) and tolerable soil loss rate as:

$$E_{water} = R*K*LS*C*P \, / \, T$$

where:

E_{water}	=	water erosion risk index (unitless)
R	=	rainfall and run-off erosivity (accounting for frequency, duration and intensity of rainfall events) (MJ mm/ha hour year, where MJ is an energy unit in megajoules)
K	=	soil erodibility (soil texture, drainage conditions, etc.) (t ha hour /ha MJ mm)
LS	=	slope length and steepness factor (unitless)
C	=	crop management factor (cropping patterns, etc.) (unitless)
P	=	conservation management factor (tillage practices, etc.) (unitless)
T	=	tolerable soil loss rate (tonnes/ha/year).

As a first step, R, K, LS and T factors for different soil types are required. These could be provided by national authorities. While at this stage C and P factors can be set to one if not available, these factors might be obtained by drawing on the development of Farm Management indicators for soils.

- *Interpretation*: This indicator combines information on the inherent vulnerability of a soil or landscape (based on physiographic and climatic properties) and information on how agricultural land is being managed. The indicator is highly relevant to policy as it reveals whether agricultural soils are being managed sustainably (from an erosion perspective) and it can identify areas at risk that would benefit from conservation efforts. Where C and P factors are not available, the value shows the level of potential risk of water erosion based on the inherent soil quality, and on physiographic and climatic conditions. In any case, a scientifically sound reference level needs to be defined.

- *Further refinement*: For USLE calculation, C and P factors need to be determined, where they are not available, possibly benefiting from the Farm Management indicators. Other approaches need to be developed and applied to address soil erosion on those aspects of land use for which USLE is not relevant, such as pastoral farming systems, terracing etc.

Risk of Soil Erosion by Wind

- *Definition*: The agricultural area subject to wind erosion (i.e. the area for which there is a risk of degradation by wind erosion above a certain reference level).

- *Method of calculation*: Index of wind erosion risk is calculated as:

$$E_{wind} = KC(V^2-\rho W^2)^{1.5}(1-R)$$

where: E_{wind} = index of wind erosion risk (treated as unitless)

K = surface roughness and aggregation (size of soil particles) factor

C = factor for soil resistance to movement by wind

V = drag velocity (wind speed at the soil surface)

ρ = variable related to the soil moisture content when erosion begins

W = surface soil moisture content

R = erosion reduction factor (crop type, crop residues, cultivation systems, etc.)

As a first step, K, C, V, ρ and W factors for different soil types are required. These could be provided by national authorities, or, when they are not available, by drawing on Canadian coefficients. At this stage the R factor can be set to zero if not available, although it might be obtained by drawing on the development of Farm Management indicators for soils.

- *Interpretation*: This indicator, which is to be treated as a unitless relative index rather than a quantitative estimate of erosion loss despite its theoretical unit of $kg/m^2/hour$, combines information on the inherent vulnerability of a soil or landscape (based on physiographic and climatic properties), and information on how agricultural land is being managed. The indicator is highly relevant to policy as it reveals whether agricultural soils are being managed sustainably (from an erosion perspective) and it can identify areas at risk that would benefit from conservation efforts. Where the R factor is not available, the value shows the level of potential risk of wind erosion based on the inherent soil quality and on physiographic and climatic conditions. In any case, a scientifically sound national reference level needs to be defined to help better interpret trends in the indicator.

- *Further refinement*: The R factor needs to be determined where not available, possibly benefiting from the Farm Management indicators.

Inherent Soil Quality

- *Definition*: Agricultural areas where there is a mismatch between the soil capability as indicated by the index of inherent soil quality and the actual or impending land use.

- *Method of calculation:* Soil maps on inherent soil qualities, which are considered to be important for each country (to be provided by national authorities), are compared with land use maps (possibly derived from using satellite imagery, but see also the report by the Group on Biodiversity, Habitats and Landscape below). A system for classifying all soils into classes needs to be developed, referring to the methodology developed by the United States at the world level, and taking into account the different data availability and various methods used in OECD Member countries. The inherent soil qualities of particular areas need only be assessed periodically, but land use information should be updated annually.

- *Interpretation:* Trends in areas on which there is a mismatch between land use and soil capability would reflect changes in the potential risk of soil degradation by various processes.

- *Refinement:* Certain conceptual issues, such as coverage and classification of different soil qualities, need to be resolved before the inherent soil quality indicator can be developed and applied to the "target" areas. The issue of map scale also needs to be resolved. The inclusion of soil organic carbon, stock and flux indicators, should also be considered, while this also links with OECD work on agricultural greenhouse gas indicators.

1.4   *Land Conservation*

Policy issues

Some agricultural management practices, including appropriate land and water management systems which regulate the water and soil sediment flow from agricultural land to downstream off-farm areas, can contribute to the prevention of floods, soil erosion and landslides, and the preservation of natural resources such as soil and groundwater. When these practices and systems are changed or land use changes from agriculture to other uses, the frequency and extent of flooding may increase. In this context, soil erosion can be exacerbated, leading to water turbidity and off-farm sediment deposition. This may harm aquatic habitats, resulting in the reduction of fish population in rivers and lakes.

The key policy objective for maintaining and enhancing the soil and water regulating functions of agriculture is to contribute to the sustainable management of environmental resources through appropriate agricultural land use and practices.

Main points discussed

It was recognised that some land conservation functions can be enhanced by certain agricultural management practices and systems, and that appropriate indicators of these functions can provide useful information for agri-environmental policy makers in most OECD Member countries. At the same time, delegates from some countries contended that the importance of these indicators is limited under their specific farming conditions. It was also recognised that conceptual and methodological refinement of these indicators would be required in order to better relate the indicators to changes in policy measures.

The relation between land conservation and other indicators was discussed, especially those concerned with soil and water. It was accepted that indicators for land conservation should focus on measuring the off-farm environmental impacts of agriculture, while soil quality and water use indicators should focus on the on-farm impacts. It was also stressed by some delegates that land conservation indicators should be linked with the issue of land abandonment (i.e. agricultural land transferred to uncultivated natural habitat, an issue covered in the set of indicators on Wildlife Habitats).

Among the proposed indicators, there was considerable interest expressed in the development of a *water buffering* indicator, which expresses the potential capacity of agriculture to control water flow and prevent flooding of downstream rivers. However, it was accepted that for the indicator to have better policy relevance a considerable number of conceptual and methodological issues regarding its construction would need to be resolved. Examples of such issues are the inclusion of factors reflecting land slope, agricultural practices and downstream human population settlements.

Although the proposed indicators for soil erosion prevention and for landslide prevention were not fully discussed, a number of delegates emphasised the importance of addressing the issue of *off-farm soil sediment flow*. It was also contended that off-farm effects on biodiversity caused by sediment flows should also be measured. For the proposed indicator on the recharge of groundwater, it was accepted that it would require more conceptual work and was accordingly not further considered at this stage.

Recommendations

Indicators for the short term

Water Buffering Capacity

- *Definition:* The quantity of water that can be stored over a short period, in the agricultural soil, as well as on agricultural land where applicable (e.g. flood storage basins) and by agricultural irrigation and drainage facilities.

- *Method of calculation:* Water Buffering Capacity (Wp) of agricultural soil is calculated as:

$$Wp \quad = \quad \sum Ai\,Pi$$

where:
- Wp = water buffering capacity (tonnes)
- A = area of land use (ha)
- i = index of agricultural land use (e.g. crop field, pasture)
- P = water retention potential (t/ha).

Data are needed with respect to (a) land area (ha) for each agricultural land use (crop field, pasture, orchard, etc. drawing on the OECD Agricultural database or national sources); (b) water retention potential (t/ha) for each agricultural land use (to be provided by national authorities; and (c) water holding capacities of various agricultural facilities such as ponds, irrigation canals, etc. (to be provided by national authorities).

- *Interpretation:* Decrease in water buffering capacity implies increased risk of flooding. Although the trends over time of the indicator values mainly reflect changes in land use, changes in land management practices or systems may also affect the indicator trend through enhancement (or impairment) of the water retention potential.

- *Further refinement:* In order to better relate this indicator to changes in policy measures, it needs to incorporate not only land use, but also other factors such as land cover, agricultural practices (cropping patterns, type of tillage, maintenance of hedgerow, etc.) and social factors (e.g. downstream human population settlements). Benchmarks also need to be determined, although one possibility is to compare the water buffering capacity of different land uses, for example uncultivated natural habitat, forestry, etc.

Further development of indicators

Off-farm Sediment Flow

- *Definition:* The quantity of soil sediments delivered to the off-farm areas from agricultural soil erosion.

- *Refinement:* This indicator measures the impact of agriculture on soil sediment flow (erosion) to off-farm areas. The risk approach proposed to develop this indicator would draw on Soil Quality indicators (see above), including the risk of water and wind erosion, possibly by linking the soil erosion rate to the quantity of soil sediments delivered to off-farm areas. Other approaches also may deserve consideration, such as state (e.g. direct measurement of suspended or accumulating solid materials in river water) and/or response (e.g. erosion control strategies) indicators of off-farm sediment flow.

2. Common and cross-cutting issues for agri-environmental indicators concerning: biodiversity, wildlife habitat and landscape

Predominant among the common and cross-cutting issues discussed by the Group were the range of issues listed below.

- The establishment of indicators in recognition of the key linkages between the areas of biodiversity, wildlife habitat and landscape, as well as the linkages to the other areas for which OECD is developing indicators (e.g. soil quality, farm management, etc.), especially because of the holistic nature of biodiversity, habitats, and landscape in particular.

- In the context of developing indicators for this Group, there was agreement on the need for some degree of national flexibility so as to enable countries to adapt overall indicator methodologies to specific agri-environmental circumstances. However, it was emphasised that this should not undermine the need for a consistent methodology in the development of indicators so that comparability between countries is still possible.

- The need to recognise the importance of the spatial resolution of developing these indicators was stressed, by taking into account the sub-national and national context as well as sub-national and national priorities for the different areas of the Group.

- In emphasising the global approach to analyse these issues, especially landscape, the need for contextual indicators was stressed, in particular, covering changes in land use and land cover, and threatening processes.

- A common theme in the discussion of biodiversity, habitat and landscape indicators was the need to develop suitable baseline/threshold values to interpret changes in these indicators.

- In the context of the OECD's "policy relevance" criterion, there was agreement on the need to refine more precisely the presumed range of policy questions that are relevant to each of the issues of agricultural biodiversity, wildlife habitat and landscape.

Against the background of these issues, and in view of the relatively recent start of work on these indicators, there was consensus that it was both premature and unproductive to seek to establish a finite and "final" set of indicators at this time. It was, however, clearly understood that practical and tangible steps had to be taken to demonstrate progress in developing an initial set of indicators.

In this vein, the Group made considerable progress in identifying the key components within each area and the indicators for the initial short-term measurement of these components, as first pragmatic steps in advancing the work. Nevertheless, significant work remains at the specific indicator level on issues related to the overall conceptual framework in which to develop these indicators, as well as to data needs, methodologies and interpretation.

It was also noted during the discussion that analysis of biodiversity, habitats and landscape is a relatively recent field of research, and that interdisciplinary research of these issues has evolved because more traditional disciplines were not comprehensive enough to address the critical concerns that have arisen. Moreover, until recently the spatial and temporal scales of examining these issues has been conducted on relatively small areas and over short periods of time. However, with the development of high-capacity computers, remote sensing data and imagery and Geographic Information Systems, there is an increasing ability to address questions of greater complexity and scale and over longer periods of time.

The Group also saw the need to take a global approach to the analysis of the impact of agriculture on biodiversity, habitats and landscape. This led to the recognition of the need to develop certain contextual indicators with respect to, for example, changes in agricultural land use and land cover, which are applicable to some degree to all three areas as well as other indicators. In addition, the possibility of constructing a matrix to reveal the various linkages and overlaps across these three areas was considered useful, while the importance of collaborating with other fora working on these issues was emphasised.

2.1 *Agricultural Land use and Land Cover Changes*

- *Definitions:*

    1. Changes between the share of land in agriculture and other uses.

    2. Changes in the share of agricultural land cover type.

- *Method of calculation:* Land use describes the functional aspects of land, characterised by some identifiable purpose or function (such as land used for agricultural, forestry or urban purposes), leading to tangible (food) or intangible products (landscape). Land cover is the description of the physical surface, which for agricultural land can encompass different crops and pasture, and the physical features such as rivers and buildings. Measuring agricultural land use/cover changes usually entails drawing data from regular agricultural censuses and, more recently, from satellite imagery, which are widely available and regularly updated. Comparing land use/cover data sets can be difficult, particularly due to varying definitions, especially as key land use/cover types (e.g. agriculture, grassland, wetlands, forestry, etc.) may vary between countries and between different users, such as statisticians and ecologists.

- *Interpretation:* These contextual indicators can be used in conjunction with the interpretation of a range of other indicators. In particular, the qualitative measures of biodiversity, habitat and landscape, but also indicators related to land conservation and socio-cultural issues (see relevant text in this report). Land use/cover indicators, for example, can help answer questions relating to the proportion of land surface annually available as habitat; the proportions of each habitat type within the agricultural area; and the spatial configuration of these habitats.

81

- *Further development of indicators:* Land use and land cover are interdependent, and the discussion in the group emphasised the importance of various driving forces in changing land use and cover. This highlighted the need to develop links between the establishment of farm management indicators and of land use/cover data sets. There is also a need to establish more clearly whether changing trends in land use/cover have positive or negative implications for biodiversity, habitat and landscape, including in the broader context of sustainable agriculture and sustainable development. This might be possible by identifying the economic, social and environmental trade-offs associated with changes in land use/cover measured, for example, in monetary terms.

2.2   *Biodiversity*

Policy issues

In many areas of the world, the influence of agriculture on the biodiversity of wildlife as well as on domesticated crops and livestock has been significant. A number of wildlife species and habitats have become more or less dependent on agriculture and their distribution, extent and evolution has been greatly influenced by it. Modern agricultural activities have led to the loss and damage of wildlife habitats and affected the number and abundance of plant and animal species, as well as the within species variation in domesticated livestock and crops.

Current trends show that intensive agricultural production is expanding in many regions. Nevertheless, less intensive systems of production remain significant. Even so, the impact of agriculture on biodiversity, either positive or negative, can depend on the system of land management and farm practices. For example, land in some areas is lost from agriculture to other uses, and conversely, semi-natural and uncultivated natural ecosystems elsewhere are cleared and transformed into farmland. In many OECD countries, particular farming systems have led to the development, over long periods, of specific biodiversity associated with these semi-natural habitats.

Main points discussed

There was agreement that, at this stage, there is not so much a need for one single, highly aggregated biodiversity indicator, such as the natural capital index, but rather a demand for more detailed underlying indicators, for example, those describing biodiversity quantity and quality. It was stressed, however, that baselines are indispensable for interpreting the state and trends in biodiversity, but that the choice of an appropriate baseline is complex and further work is required in this area. For policy makers, however, a practical baseline would have to be calculated, based on a specific period over the past few decades.

There was also recognition that as a useful first approximation to a complex and poorly understood issue, agriculture's impact on biodiversity could be conceived in terms of the categories of: agricultural genetic species diversity; wild species that are important in supporting agricultural production activities (e.g. bees); and wild species that are affected by and/or dependent on agriculture. Even so, indicators of biodiversity of wild life species that are important in supporting agriculture, are still in an early phase of development and require further research. The Group also agreed that work on indicators of agriculture's impact on biodiversity would benefit from closer collaboration with the OECD Group on the Economic Aspects of Biodiversity, in the case of genetic diversity with activities underway in FAO, and more broadly with the work of the Convention on Biological Diversity.

The Group concluded that *"quantity" indicators* of biodiversity, which mainly concern the extent and type of coverage of agricultural land, could be established by drawing on the on-going work on indicators for wildlife habitats, and on those concerning agricultural land use and land cover. For developing such "quantity" indicators the question of spatial variation and technical costs were raised as being problematic for larger countries. There was also consensus that varying spatial scales should be used, in order to reflect country differences in agricultural ecosystems.

It was agreed that *"quality" indicators* of biodiversity should consist of several variables, including species and habitat indicators. Nevertheless, concerns were identified in connection with the feasibility of some quality indicators, especially as there currently exists no standardised methodology for choosing common species or taxonomic groups across countries.

Since comprehensive species data are unavailable for most countries, there was a general view that the state of and trend in the populations of certain indicative species (e.g. birds) and/or "threatening" processes (e.g. rabbit invasions), could serve as a useful proxy of biodiversity quality. Even so, there was agreement that a pragmatic approach would be needed to choose indicative (endemic) species, or groups of species, that are important to the functioning of particular agricultural ecosystems.

An important factor pertinent to the development of any of these quality indicators is that the choice of species and/or threatening process will have to be made by the countries concerned, in order to reflect differences in national ecosystems and priorities. At the same time, it is important that a consistent approach across countries be adopted when choosing indicative species for comparative purposes.

Because of the enormous difficulties in measuring the inherently complex and multidimensional nature of biodiversity quality, a number of other indicators serving as approximations were considered. Among the more promising discussed were the percentage of semi-natural and uncultivated natural habitats within the total agricultural area (drawing on the land use/cover indicators), and a typology of agricultural systems and farm practices based on their impact on biodiversity.

Recommendations

Indicators for the short term

Genetic Diversity of Domesticated Livestock and Crops

- *Definitions:*

    1. Change in the sum of all recognised and utilised varieties of domesticated livestock and crops.

    2. Change in the share of different livestock and crop varieties in the total population or in total livestock and crop production.

- *Method of calculation:* These indicators require data covering livestock/crop species, and total livestock/crop populations and production. It will be necessary to establish a system by which to collect species data, with genebanks and national breeding organisations providing a starting point.

- *Interpretation:* The prevention of the erosion of genetic diversity is important, as genetic material loss is generally irreversible. The baseline from which this loss should be measured is yet to be determined. Indicator 2 could serve as a short-term approximation of genetic loss in agricultural production systems.

- *Further development of indicators:* It would be useful to examine the distinction between species commonly used for intensive production and those used in extensive systems. In addition, it is necessary to determine a suitable baseline from which to interpret changes in the genetic diversity of agricultural livestock and crops.

Wildlife Species Diversity related to Agriculture

- *Definitions:*

  A. Quality

  1. Appropriate key species indicators for each agro-ecosystem.

  2. Key threatening processes that can damage agricultural production activity.

  3. Proportion of semi-natural and uncultivated natural habitats on agricultural land.

  B. Quantity

  4. The extent of changes in the agricultural area and type of land cover (this indicator would draw from the wildlife habitat and land use/cover indicators).

- *Method of calculation:* For indicators 1 and 2, the choice of species (possibly surrogate species, taxonomic groups or Red List — endangered species) and threatening processes to agricultural production (such as invasion of pests) would be left to individual countries. Further reflection on defining "appropriate agro-ecosystem" would also be required. The possibility of returning to standardised approaches could be explored once countries have identified surrogate species, taxonomic groups or threatening processes. The choice of species and/or threatening processes, as defined in these indicators, may need to reflect differences in national ecosystems and priorities. Indicators 3 and 4 overlap with the indicators of wildlife habitat, and the indicators identified here will draw from this work. For developing "quantity" indicators of biodiversity the question of spatial variation and technical costs were raised as being problematic for larger countries.

- *Interpretation:* A reduction in key species populations (indicator 1) or an increase in threatening processes (indicator 2, for example, an expansion in rodent animal numbers) can be seen as diminishing wildlife species diversity. However, the interpretation of these indicators is not straightforward, and care will be required in relating species reductions or increases to agriculture, where other external factors, such as changes in weather or populations of natural living organisms, may have an important influence. The reduction in habitat area (indicators 3 and 4) would also be viewed as a threat to wildlife species diversity. As with genetic diversity the baseline from which to interpret changes in wildlife species diversity is yet to be determined.

- *Further development of indicators:* Baselines from which to interpret changes in biodiversity, are indispensable for valuing the state and trends in biodiversity. Since the choice of an appropriate baseline is complex, however, further work is required. There is also a need for further work with respect to measuring changes in the numbers of endangered species related to agricultural ecosystems, and to the impact on biodiversity of different agricultural systems and farm practices. A feasible approach to the latter might be through using a set of weights specifying how each different system or practice affects biodiversity. The effects caused by off-farm soil sediment flow on biodiversity might also be further examined in conjunction with the land conservation indicators.

2.3   *Wildlife Habitat*

Policy issues

The maintenance of certain wildlife habitats on agricultural land, semi-natural and uncultivated natural habitats in particular, contributes towards achieving a number of environmental objectives, including those associated with international obligations for the protection of rare or scarce wildlife habitats and their associated species. Wildlife habitats can serve as a buffer-zone and protect natural resources, for instance by reducing soil erosion. Certain wildlife habitats also have the effect of maintaining the ecosystem's biological processes of self-regulation by improving the survival chances of beneficial species (natural enemies of crop pests) thus allowing, for example, a reduction of the use of pesticides in agriculture. In addition, certain wildlife habitats in rural areas can help improve the aesthetic quality of the landscape and increase its recreational value, for both the population in general and, in certain cases, for tourism.

According to the definition given by the International Convention on Biological Diversity, the conservation of the diversity of species and agricultural ecosystems — which are the two major objectives of the maintenance of wildlife habitats — represent two of the key categories of biodiversity conservation. The concept of habitat is on a lower level than the one of "ecosystem", which is itself on a lower level than a bioregion. An ecosystem includes both living (biocenose) and non-living elements (biotops), and is usually in contact with one of the five other ecosystems surrounding the agricultural system, including: forest, aquatic, steppe, rocky and urban ecosystems. Hence, habitat, as a small part of an ecosystem, includes both biocenose and biotop aspects, but is limited to an area where a certain number of ecological factors are quite homogeneous. For example, a field of wheat or a meadow or a hedge as habitat, is considered as being part of an agricultural ecosystem.

Main points discussed

The three proposed categories of indicators covering agricultural habitats were broadly endorsed as covering habitats that are either intensively farmed, semi-natural or uncultivated natural habitats. There was considerable discussion over whether indicators of biodiversity and wildlife habitat ought to be merged, but the consensus was to maintain separate indicators while recognising both their specific characteristics as well as any overlap between these and indicators for biodiversity and landscape.

Recommendations

Indicators for the short term

Intensively Farmed Agricultural Habitats

- *Definition:* The share of each crop in the agricultural area.

- *Method of calculation:* The percentage of the agricultural area covered by each crop type, covering the main marketed food, fibre and industrial crops. This indicator is the same as the indicator concerning land cover (see Section 2.1).

- *Interpretation:* Each type of crop has, to a greater of lesser extent, a favourable influence on wildlife, although the relationship between specific crops and wildlife species has yet to be firmly established.

Semi-natural Agricultural Habitats

- *Definition:* The share of the agricultural area covered by semi-natural agricultural habitats.

- *Method of calculation:* The percentage share of the agricultural area covered by semi-natural habitats, with a precise list and definition of the semi-natural habitats to be covered and calculations made of the relevant areas covered by these habitats. Semi-natural agricultural habitats can be broadly defined as habitats not subject to intensive farming methods and on which the use of fertilizers and pesticides is absent or restricted. This indicator is closely connected to the indicator of the Wildlife Species Diversity indicator (see Section 2.2 above).

- *Interpretation:* The larger the area covered by semi-natural agricultural habitats in relation to the agricultural area the more beneficial are the effects on wildlife, as in general these habitats harbour a much greater variety and abundance of species than on intensively farmed agricultural areas.

Uncultivated Natural Habitats

- i) *Definitions:*

  1. Area of wetland transformed into agricultural area.

  2. Area of aquatic ecosystems transformed into agricultural area.

  3. Area of natural forest transformed into agricultural area.

- *Method of calculation:* The area of wetlands, aquatic ecosystems (i.e. rivers, ponds or lakes) and natural forests transformed annually into agricultural land (for rivers the unit of measurement is length). The term "natural forest" may encompass "secondary" forests, which are forests exploited by man in a sustainable manner and in which the physical condition and diversity closely resemble the natural state, having developed over a long time period.

- *Interpretation:* These indicators reveal whether or not habitats, which are important for many species, are being preserved (e.g. wetlands, aquatic ecosystems, natural forest). The transformation of these habitats for agricultural use is generally linked with a high diminution of their value with regard to the wildlife in the area concerned, especially as these habitats may have taken several thousand years to evolve.

- ii) *Definition:*

  4. Area of agriculture re-converted into aquatic ecosystems.

- *Method of calculation:* This indicator measures the area of agricultural land reconverted or transformed back into an aquatic ecosystem.

- *Interpretation:* The reconversion or transformation of agricultural land back into an aquatic ecosystem can usually be interpreted positively.

Further development of indicators

A key prerequisite before measuring these indicators may be the establishment across OECD countries of common definitions of the major types of habitat identified here. This is particularly the case with respect to the classification of the three main groups: intensive, semi-natural and natural habitats. In some cases wildlife habitat indicators overlap and/or could draw on the agricultural land cover and land use indicators (for example, the area of agricultural land changed into other uses, such as uncultivated natural habitats). The possibility of analysing the various habitat indicators in one set could be explored. Also the feasibility of the extension of wildlife habitat indicators to include habitat heterogeneity (average size of habitats) and variability (number of habitat types per monitoring area) might be examined. The linkages between different farming practices and their impacts on biodiversity and wildlife habitats also need to be further explored (see also Section 3.2 on farm management, footnote to Figure 4, and reference to organic farming, Box 1).

2.4   *Landscape*

Policy issues

At the international level the issue of landscape has only been addressed relatively recently by researchers and policy makers, although rapid developments to better understand the issue are in progress. This is, in particular, due to the increasing policy focus on agricultural landscapes in many OECD countries.

Despite the variety of national and professional interpretations, a comparison between various landscape definitions shows a series of common landscape elements related to spatially differentiated:

- *land characteristics*, including the natural biophysical features (e.g. topography, geomorphology); the environmental appearance (e.g. ecosystems, habitats); and land type features of the landscape (e.g. land cover and land use);

- *cultural features*, including cultural amenities and aesthetic features and values related not only to landscape, such as hedges, the number of "historic" monuments or sites, walls, and tree patterns, but also to the type and amount of recreational facilities directly attributable to the landscape;

- *management functions*, in particular the public and private initiatives to maintain landscape quality, but also including forms of farm management systems and practices, financial resources available to farmers, and aspects of socio-cultural issues (rural viability). These include, for example, demographic trends of new entrants into farming which give rise to, and effect changes in, certain agricultural landscape types. The management functions of landscape depict the joint production aspects of agriculture, that is certain farming systems and practices give rise to not only commodity production but also the production of landscape.

Landscapes can be identified as individual spatial units where region-specific components and processes reflect the interaction between the land characteristics, cultural features and management functions. Moreover, the underlying human and natural processes are subject to change and evolution, hence, landscapes are dynamic systems. For the purpose of the OECD work the main focus is on landscapes that have evolved from agricultural land use practices.

An analysis of the existing policy background demonstrates that landscape planning and regulation at the national level is underway in many countries. At the international level, landscapes are attracting increasing policy attention. This is documented by a number of new environmental programmes and policies, including a number of European initiatives, and amendments to the Convention on Biological Diversity. A recent review among certain OECD countries, provided in the background paper on landscape, has demonstrated that the various approaches towards landscape assessment are relatively similar in terms of their scope, objectives and focus on landscapes.

Main points discussed

In developing indicators to address the landscape issue, a number of points were agreed.

- Landscape encompasses a basic spatially defined unit area which incorporates elements from many other agri-environmental areas. In this context it is important to recognise the linkages with biodiversity and habitats, in particular, but also with, for example, soil quality and farm management. The landscape issue also has linkages with a number of other areas of OECD work, particularly those concerning territorial development and rural amenity.

- The indicators selected by the Group to encapsulate each of the three elements of landscape should be viewed as composite indicators. Moreover, the interlinkages between the indicators also needs to be stressed, for example, indicators of land cover and land use to measure the land characteristics of landscape, also encompass a combination of cultural and management elements.

The need for a global approach to develop component indicators for analysing landscape was agreed. Moreover, as the landscape issue is one of the least developed areas for which indicators are being established, the work should develop in stages so as to encompass the complexity of the issue and facilitate the combination of information from different statistical databases.

There was considerable discussion concerning the possibility of developing a system of landscape types or typology, especially as a number of countries have already established such systems. One possibility is to define landscape in terms of the most important processes by which agriculture affects landscape, including:

- Expansion-Withdrawal.

- Intensification-Extensification.

- Concentration-Marginalisation.

These processes are usually linked, as in the case, for example, of agricultural marginalisation and land withdrawal from agriculture. They also operate on all scales from the farm to the national level. Such an approach enables landscape indicator development to focus on some of the most policy-relevant and significant impacts of agriculture on landscape. A typology of landscapes also allows the impact of changes in these processes to be assessed indirectly by indicators, and the results assessed for each national situation depending on which process is viewed as having a positive or negative impact. Hence, landscape typologies can be important in establishing the definition of baselines or threshold levels for interpreting positive or negative changes in landscape indicators. Landscape typologies could also be used to follow the dynamic development of landscapes through measuring the change of landscape types.

Recommendations

Indicators for the short term

Land Characteristics of Agricultural Landscape

- *Definitions:*

    1. Natural features, covering, for example, the land's slope, elevation, soil type, etc.;

    2. Environmental appearance, including the landscape ecosystems and habitat types;

    3. Land type features, including changes in agricultural land use and land cover type.

- *Method of calculation and interpretation:* These indicators largely draw directly from other areas described in this report, including indicators of soil quality, wildlife habitat and land use.

Cultural Features of Agricultural Landscape

- *Definition:* Key indicative cultural features.

- *Method of calculation:* This indicator relates to key cultural features of landscapes, and would be determined according to different national situations and priorities, as well as being associated with land use. It would, for example, involve measuring changes in the length of linear landscape features such as hedges and dry stone walls, and in spatial features such as the historic and cultural monuments and sites on agricultural land.

- *Interpretation:* Identification of the key cultural features are considered important to conserve or improve as key elements in the current state of the landscape. A downward trend in such indicators would be considered an undesirable development. An example might be the reduction in the length of hedgerows. The interpretation of trends in these indicators, however, ultimately depends on what constitutes a valued landscape which, in turn, is a reflection of cultural preferences and aesthetic values. At present there are no systematic national efforts to measure these preferences and values, although some countries are exploring the use of public surveys combined with various valuation techniques to measure public preferences in monetary terms.

Management Functions of Agricultural Landscape

- *Definition:* The share of agricultural land under public and private commitment to landscape maintenance and enhancement.

- *Method of calculation:* The share of farms, or land area, covered by public/private schemes or plans that provide a commitment to landscape maintenance and enhancement, calculated annually as a percentage share of the total number (or area) of farms (agricultural land).

- *Interpretation:* Identification of the public/private initiatives (or responses) in farm management practices and systems which act to maintain and enhance landscape features, by maintaining and/or impairing, particular land use and land cover patterns and the cultural features of the landscape. It is assumed that the greater the number (or area) of farms covered by public and private landscape initiatives, the better will be the maintenance of, or improvement in, the quality of the landscape.

Further development of indicators

There are a number of key elements in further developing landscape indicators, including:

- Drawing on other indicators, in particular indicators covering changes in agricultural land use and land cover, biodiversity, wildlife habitats, and farm management.

- Identifying more precisely the methodological linkages with other indicators, so that this information can be combined and integrated into the development of a coherent set of landscape indicators.

- Developing methods to measure the value society places on agricultural landscapes, in order to assist policy makers in determining the costs and benefits stemming from conservation. Public surveys and use of monetary valuation techniques to measure societal landscape preferences, underway in some countries, provide a useful starting point to advance the work on valuing landscape. This is also relevant where society (or communities at the sub-national level) attach important aesthetic and spiritual values to the landscape, for example, some communities prefer an "open" agricultural landscape, while others attach spiritual values to the land based on particular beliefs and historical antecedence. The monetary valuation of public preferences for landscape, however, is attended by various technical problems which should not be underestimated, in particular the difficulty of cross-country comparisons.

- Establishing relevant baselines/threshold values to facilitate the interpretation of negative/positive trends in landscape indicators.

3. Common and cross-cutting issues for agri-environmental indicators concerning: farm management, farm financial resources and socio-cultural issues (rural viability)

The work on agri-environmental indicators captures the broader economic and social elements of sustainable agriculture, in addition to the environment. The Group agreed on four concepts for indicators covering this group:

- Farm management capacity (covering the institutional aspects of agriculture).

- On-farm management practices.

- Farm financial resources.

- Socio-cultural issues (rural viability).

The first two concepts cover the environmental dimension of sustainable agriculture, the third is related to the *economic* dimension, while the last covers *rural viability* (or the social dimension of sustainable agriculture). The linkages between these different concepts in the context of environmental farm management plans and their impact on the environment are illustrated in Figure 3.

The framework proposed in this report allows for the further refinement of relevant indicators and for flexibility to accommodate the diversity in farming systems and ecological and environmental conditions across countries. The Group emphasised the need to draw on other related work in the OECD, such as rural development and structural adjustment.

Some concepts will need further evaluation and/or development (e.g. social capital). Many of the proposed indicators are surrogate indicators or proxies, reflecting in particular the difficulty of obtaining consistent data on farm management, financial resources and socio-cultural issues (rural viability).

Figure 3. Linkages between the indicator concepts

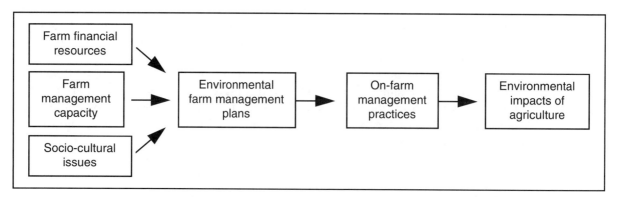

While basic quantitative data, such as land use, farm accounts and pesticide and nutrient use are available in many countries or regions, information on factors relating these data to the environmental impacts of agricultural activities are often missing or at an early stage of development. The emphasis should therefore be on developing ways to measure the outcomes of farm management on the environment.

Links between indicators should be clearly identified. Indicator areas that are closely related should be viewed together. For example, farm financial resources are closely related to farm management and socio-cultural issues (rural viability). The Group also identified some contextual indicators, such as changes in land use, number of farms and agricultural population, that are closely related to all three areas, as well as to other indicators.

3.1   *Farm Management*

Policy issues

Farm management practices have a direct impact on the environment, both on and off the farm. Decision makers can benefit from information on the nature of current farm management practices used, how these practices affect the environment, and how they compare with recommended (or legislated) practices and standards. Farm management decisions are in turn influenced by policy actions, including financial support and investments in research, education, extension, etc.

The Group recommended measuring two key concepts of farm management, both oriented towards the achievement of environmentally sustainable agriculture:

- farm management capacity (based on data at the aggregated agriculture sector level); and

- farm management practices (based on data measured at the farm level, but aggregated to the national level).

Indicators on *farm management capacity* cover the investment by the wider society in the capacity of the agricultural sector to build and transfer knowledge to improve on-farm management practices, that will lead toward a more environmentally sustainable agriculture. This investment covers a broad range of elements, but in particular investment into farmer education, research and the development of appropriate institutions and standards, to encourage environmental farm management practices and farming systems.

Indicators on *farm management practices* encompass the overall trends of farming methods. They address whole farm management as well as various aspects of farm management, such as nutrient management, pest management, soil and land management and irrigation management, which have significant effects on the environment.

Environmental conditions and farming systems vary within and across OECD countries and, consequently, optimal farm management practices vary from one region to another. For example, a detailed nutrient management plan is not a priority in areas without nutrient surplus or leaching problems. Nor is there a need to change pest control practices if pesticide use is already at a low level for climatic or other reasons. Thus, identifying and developing a standard set of indicators on farm management practices across the OECD is not straightforward. A matrix of farm management practices which allows the diversity of country situations to be reflected, was proposed as a tool to accommodate this variability.

Discussion distinguished between the:

- advisory and information inputs into farm decision making;

- formulation of plans, strategies and schemes for the farm; and

- environmental consequences of farming activities and practices.

Main points discussed

The Group agreed that:

- Farm management indicators can provide an early indication of changes in direction, either positive or negative, sometimes well before the actual environmental impacts can be measured by other agri-environmental indicators, for example, soil and water quality.

- Farm management indicators can also serve as a proxy for "state" indicators where the latter are difficult or more costly to monitor. Measuring farming practices is often more practical and cheaper than measuring actual changes in the environment.

- These indicators can be useful in assessing linkages between farm practices and the state of the environment. Monitoring the trends in management practice indicators alongside appropriate "state" indicators, will allow policy makers to evaluate directly the success of policies aimed at environmental improvement. Indicators on farm management are closely linked to other indicators, such as nutrient use, pesticide use, water and soil quality.

- The indicators are important because they are driving forces (and, in some cases, responses) which relate directly to agriculture, whereas changes in state indicators, such as water quality, may result from changes arising outside of agriculture, such as industrial pollution.

There was discussion on what is the optimal measurement point (from an indicator perspective) for assessing a given system or management practice. The Group agreed that it is better to measure actions or practices undertaken by farmers, rather than intentions, especially as it is often difficult to measure environmental outcomes (i.e. measure what farmers actually do, not what they say they intend to do).

Recommendations

Farm management capacity

Indicators for the short term

Standards for Environmental Farm Management Practices

- *Definition:* Number of established national and/or sub-national environmental farm management standards, regulations, codes of practice, etc.

- *Method of calculation:* This qualitative indicator is an inventory or descriptive listing of environmental farm management standards, etc., established by public agencies, but also voluntary, agriculture industry self-generated standards, etc. The Group recommended this indicator as a starting point for measuring environmentally sound management practices. This information will be included in a more refined form in the matrix on environmental farm management practices. Standards, regulations, codes of practice, etc., do not change very often, so there would be no need to update the indicator annually. National authorities could compile this data from records of government agencies, professional organisations, industry, farm surveys, etc.

- *Interpretation:* The establishment of professional standards and specifications indicates the intention to develop farm practices that are environmentally sound, reliable and valid. An increasing trend would potentially imply a greater desire (or intention) to move towards a more sustainable agriculture. The role of sub-national jurisdictions may be greater than that of national governments in developing and enforcing regulations, codes of practice, etc.

- *Further refinement:* The existence of standards and specifications does not necessarily mean that the actual practices have a firm scientific basis or are implemented, monitored and effective as well as efficient. Hence this indicator measuring "intent" would need to be linked closely with those indicators on "actual" implementation of farm management practices.

Expenditure on Agri-environmental Research

- *Definition:* Expenditure on agri-environmental research as a percentage of total agricultural research expenditure.

- *Method of calculation:* Annual expenditure on agri-environmental research conducted and funded by government, private sector, non-profit institutions and universities on agri-environmental issues. Data would be derived from national authorities, consisting of annual reviews of expenditure, surveys, public records, etc. The OECD Research and Development (R&D) database contains data on R&D expenditure of different sectors, but the breakdown and the level of detail, may not yet be sufficient to identify expenditure on agri-environmental research.

- *Interpretation:* Research and development are clearly bound up with rapid technological changes, leading potentially to both positive and negative impacts on agriculture and the environment. Trends in research funding reflect government priorities. Increasing appropriate public and private research funding can contribute to sustainable agriculture.

- *Further refinement:* An important refinement of this indicator will be to define "agri-environmental research". For example, should research on genetically modified organisms be excluded? Research funding levels do not necessarily reveal the relevance of the research to agri-environmental issues, whether the outcomes support sustainable agriculture objectives, what is the quality of the research, or whether farmers will accept and apply the research findings. The indicator could be further developed to reflect at least some of these aspects, particularly by linking to farm management practice indicators.

Educational Level of Farmers

- *Definition:* Average educational attainment of farmers, presented as the share of farmers attaining different levels of education or years of education.

- *Method of calculation:* A possible way of showing the annual breakdown of educational attainment level by farmers draws on the EUROSTAT Farm Structure Surveys.[4]

  – Only practical experience: Experience acquired through practical work on an agricultural holding.

  – Basic training: Any training courses completed at a general agricultural college and/or an institution specialising in related subjects (including horticulture, veterinary science, agricultural technology and associated subjects). A completed agricultural apprenticeship is regarded as basic training.

  – Full agricultural training: Any training course continuing for at least two years after the end of compulsory education and completed at an agricultural college, university or other institute of higher education in agriculture, horticulture, veterinary science, agricultural technology or an associated subject.

- *Interpretation:* There is evidence to show a positive correlation between education level and effective farm management and timely adoption of environmentally sound management practices.

- *Further refinement:* This indicator could be developed in the longer term to include the share of farmers trained in environmental farm management practices, etc. In addition,

---

4. See, for example, EUROSTAT (1996), *Farm structure: Methodology of Community Surveys*, Section B/03, Luxembourg. Other data sources that could be drawn on to develop this indicator include OECD's data on education levels by economic sectors. Data on the education level of the agricultural population were published, for example, in *Farm Employment and Economic Adjustment in OECD Countries* (OECD, 1994).

further work will be necessary to more clearly establish the link between educational attainment levels and adoption of certain farming practices.

Further development of indicators

Ratio of Agricultural Advisers

- *Definition:* Number of public and private agricultural advisers trained in environmental management practices per farmer.

  The indicator could be further developed to cover adoption-diffusion methods being developed or in place for environmental farm management practices (information flow from farmer to farmer, including demonstration farms run by farmers, etc.). Data availability would be a problem especially in the short term.

Environmental farm management practices

Indicators for the short term

Matrix of Environmental Farm Management Practices

- *Definition:* A matrix on environmental farm management practices is a tool through which the diversity of environmental conditions and optimal farming practices can be recognised and organised.

- *Method of calculation:* The matrix includes an issue substructure (nutrients, soil, pesticides, water, etc.) and specified management practices under each, with countries reporting on the level of adoption or "actual" use of those practices most relevant to their specific national and regional situations. Figure 4 below illustrates the structure of the matrix, with Boxes 1-5 describing some farm management practices that could be included in the short term. Other relevant practices, such as farm practices to protect biodiversity and habitat, could be added.

  The focus in the short term should be on measuring specific management practices; both the share of farms (or land area) using the practice and its implementation.

- *Further refinement:* The Implementation Index, see Figure 4, could be used to measure the extent to which environmental farm management practices are actually used by farmers. It would be a way to express the results of the matrix in a comprehensive manner for a given country. The Implementation Index (II) could be calculated as follows:

$$II = \sum_{i=1}^{n} \frac{\left[(\text{adoption rate of practice i at time 2}) - (\text{adoption rate of practice i at time 1})\right]}{\text{Total number of practices monitored}}$$

where: i is a management practice; i = 1,2,...n and

$$\text{adoption rate of practice i} = \frac{\text{Number of farmers using practice i}}{\text{Total number of farmers}}.$$

The adoption rate of practice i would thus be a sub-index showing the change over time in the ratio for each practice in a country or agro-ecological region.

Agricultural Census and farm surveys could be used as data sources: farmers could be asked about their use and adoption of specific practices through these. In addition, satellite imagery or GIS can be used to measure some practices.

Figure 4. Indicators of sustainable farm management practices

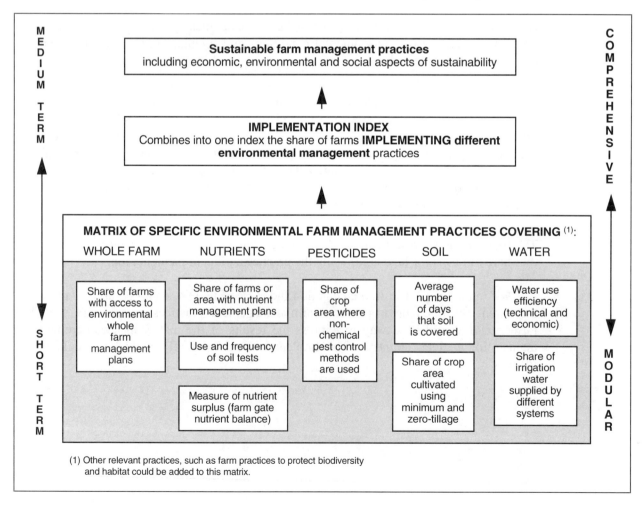

Box 1. Whole farm management

Definition:

Share of farms with environmental whole farm management plans.

Method of calculation:

The number or area of farms with access to a whole farm management plan (e.g. ISO 14000 certification), expressed as percentage of total number (or area) of farms. This can include farms under organic methods of production. Data drawn annually from national administrations.

Interpretation:

An indicator of farmer awareness of environmental issues: farmers are being encouraged to develop whole farm plans by reviewing potential environmental problem areas on the farm and developing action plans to address issues that do not meet standards. As the implementation of the plan is the farmers' responsibility this is not necessarily a precise indicator of "actual" implementation, rather an indicator of intent (unless the particular practice or plan is compulsory (e.g. farm waste disposal regulations) or obligatory as part of receiving payments under particular programme (e.g. land set aside). When implemented, environmental farm plans assure that best management practices are applied to all aspects or the whole of the farm. The greater the number or area of farms covered by those plans, the better this is for the environment, through superior farmer knowledge and awareness and best management practices.

Further refinement:

There is no consistent definition of what constitutes an environmental whole farm management plan (or organic agriculture) in the context of sustainable agriculture. The existence of an environmental whole farm plan (or organic farming) does not indicate the quality of the plan or whether the plan is implemented. These aspects should be included in the future.

Box 2. Nutrient management

Definition: Share of farms with nutrient management plans	*Definition:* Nutrient surplus measured by farm gate nutrient balance	*Definition:* Use and frequency of soil tests
Method of calculation: Share of farms, or area, with a plan to balance the inputs of nitrogen (N) and phosphorus (P) with crop needs of these nutrients (using average content of N and P by analysis or from literature).	*Method of calculation:* Share of farms, or area of farms, having significant nutrient surplus. Calculated as balance of N and P purchased in fertilizer and feed, plus fixation of atmospheric N by legumes minus amount of N and P in product sold from farm.	*Method of calculation:* Number of farms conducting soil tests and their frequency by ecoregion or country.
Annual data from national authorities (from national surveys, or sampling of representative farms).	Annual data to be provided by national authorities. Detailed farm-level data on N and P in fertilizers used, N fixed by area of legume crops (using rates from literature), and exported in grains, milk, meat, eggs, etc. (using average content of N and P by analysis or from literature).	Annual data from national authorities on field or farm numbers reported by soil test laboratories.
Interpretation: Widespread use of nutrient management plans suggests sound nutrient management and requires a good understanding of the economics of different nutrient sources and handling options (i.e. understanding of crop needs and nutrient requirements at different growth stages, in order to match nutrient applications efficiently to absorption by the crop roots).	*Interpretation:* Positive value indicates surplus of nutrients that may harm environment. Negative value indicates soil "mining", which may result in long-term soil quality problem. Target should be zero or insignificant surplus.	*Interpretation:* Greater the number of soil tests, greater the likelihood that applications rates are matched to crop needs. Target is one or more soil tests per farm per year. An indicator of interest and awareness, even if recommendations are not always followed.
Further refinement: The existence of a nutrient management plan does not necessarily mean that the plan is followed, thus it is the implementation of the plan that should be measured.	*Further refinement:* This indicator is a refinement of the soil surface balance, which is the national level indicator for *Nutrient use.* At present, the farm gate balance could only be calculated for a small sample of representative farms in each ecoregion or country.	*Further refinement:* Some soil tests do not include N.

Box 3. Pest management

Definition:
Use of non-chemical pest control methods.

Method of calculation:
Share of annual crop area where non-chemical pest control methods are used, expressed as percentage of total crop area.

Annual data from national authorities on cropland area treated with biological control agents (e.g. parasitic organisms for control of insect pests) or where weed control is achieved with tillage or non-chemical methods (e.g. ploughdown of allelopathic residues, that is plants whose roots and residues can suppress the growth of many other plants, including weeds).

Interpretation:
Non-chemical pest control methods can be used to manage pest pressures without affecting the farm economic health. Use of these alternative practices reduces pesticide use, and the risks to man and the environment.

Further refinement:
The definitions of practices need to be harmonised to improve international comparability. The indicator could also be developed to include, for example, Integrated Pest Management (IPM). Data availability on areas where both chemical and non-chemical methods are used in parallel, including IPM, could be improved.

Box 4. Soil and land management

Definition:
Number of days per year that soil is covered.

Definition:
Use of reduced and zero-tillage and other best land management practices including crop rotations.

Method of calculation:
The indicator can be sub-divided by the percentage of soil cover provided by vegetation and crop residues.

Method of calculation:
Share of crop area cultivated using minimum and zero-tillage practices, crop rotations, grassed waterways, contour strip cropping, etc.

Annual data from national authorities on type of crop, planting, tillage and harvesting dates, residues remaining after harvesting, residues remaining after each tillage operation.

Data from national authorities from census questionnaires or from sample surveys of representative farms.

Interpretation:
Plant and crop residue cover protects soils from erosion, reduces run-off of nutrients and pesticides and provides habitat for biodiversity. Greater the cumulative soil cover, the greater the protection from soil erosion, compaction and run-off, and the greater the contribution to biodiversity.

Interpretation:
Indicator of the use of best management practices for crop production to minimise soil erosion, etc. The higher the adoption of such practices on land areas at risk that require them, the lower the risks of soil erosion, etc.

Further refinement:
Defining more clearly the relative efficiency of different soil cover types in terms of nutrient and pesticide run-off, etc.

Further refinement:
Defining more clearly the relative efficiency of different practices in reducing soil erosion, etc.

Box 5. Water/irrigation management

Definition: —Water Use (Technical) Efficiency: For selected irrigated crops, the mass of agricultural produce (tonnes) per unit of the volume of irrigation water consumed (the volume of water, in megalitres, diverted or extracted for irrigation less return flows).
—Water Use (Economic) Efficiency:
For all irrigated crops, the monetary value of agricultural production per unit irrigation volume consumed (the volume of water, in megalitres, diverted or extracted for irrigation less return flows).

Definition: Irrigation delivery systems

Method of calculation: (For more detailed discussion, see the section on Water Use Indicators of Chair's Summary Report of Group 1).

Method of calculation: Share of irrigated agricultural area under different irrigation systems (e.g. flooding, high pressure rainguns, low pressure sprinklers and by drip-emitters).

Annual data on the volume of consumed water for irrigated crops, as well as yields and value of irrigated crops, need to be collected for a certain period of time. In order to remove the annual fluctuations caused by changes in climatic conditions and commodity prices, interpretation needs to focus on pluri annual trends which may reflect changes in selection of crop type irrigated, irrigation practices and trends in crop productivity.

Annual data from national authorities collected by farmer questionnaires or from irrigation district records.

Interpretation: Indicates technical (economic) water use efficiency. The greater the percentage of irrigation water applied by high-efficiency application (e.g. drip-emitters) compared with low- efficiency systems (e.g. flooding), the less the amount of water wasted and the lower the risk of adverse environmental effects. The target is to apply water by the most technically and economically-viable high efficiency system.

Further refinement Need to assess the relative efficiency of different irrigation systems (flood, trench, sprinkler, dripper, etc.). The efficiency of storage, diversion and delivery systems (i.e. water lost in conveyance systems) could be evaluated and price distortions by agricultural policies excluded.

Further refinement Need to clearly define the hierarchy of technical (economic) efficiency of different irrigation systems.

3.2 Farm Financial Resources

Policy issues

Indicators on farm financial resources illustrate the relationship between financial resources and farm management practices. The latter contribute to the sustainable use of environmental resources on and off the farm. The effects of these financial *driving forces* can only be assessed with a better understanding of their role within the wide range of factors that influence farmers' decisions and management behaviour.

Financial resources available to the farm may impact on: the ability to farm; the type, level and intensity of input use and production; the ability to acquire new technologies; the extent of adoption of environmental production methods, including farmers' attitude towards environmental risks; rates of structural adjustment including farm amalgamation, exit and entry; and the pressures for policy interventions.

The timing, certainty and level of financial resource flows can affect farmer ability and behaviour with respect to the environment, especially in countries where farmers finance environmental protection themselves. The financial resource available to the farm may also affect farmers' incentive to adopt practices that promote sustainable agriculture, to the extent that farmers have sufficient knowledge, and the environmental costs are internalised in the financial performance of the farmer.

In order to assess sustainability, a farm using natural capital stock needs to take into account the cost or return that reflects changes in the resource capital on and off the farm. This sustainable cost approach involves adjusting the returns for the cost of environmental effects, depletion of natural resource stocks, etc.

Main points discussed

The Group discussed at length whether indicators on farm financial resources should be considered as agri-environmental indicators, since they are related to the financial health of the farm and not directly to the environment. However, the Group agreed that these indicators are relevant for assessing economic sustainability and sustainable agriculture: farmers need to earn a sufficient income to enable future investment and household consumption which allows the environment to be taken into account in their management decisions.

The Group recognised the necessity to improve the understanding of the complexity of linkages and feedbacks between the level and variation in farm financial resources and environmental impacts. In addition, cause and effects of such impacts and the responses by farmers, policy makers and society to changes in agri-environmental conditions also require further study (Figure 3). These relationships may vary considerably both within and among countries. The farmers' skill and ability to manage financial resources; the form of pluriactivity; the size and ownership structure of farms; the physical agro-ecosystem; and the policy, economic and socio-cultural context in which farms operate, all influence the relationship between farm financial resources and the environment.

Indicators for the short term

Public and Private Agri-environmental Expenditure

- *Definition:* Public and private expenditure on agri-environmental goods, services and conservation (both investment and current expenditure).

- *Method of calculation:* The indicator, measured annually in current and real terms per farm, should take account of: public expenditure on agri-environmental programmes/measures; existing financial support programmes for farmers adopting environmental farm management practices (cost-share programmes); private financial resources from farmers directed to environment, etc. Trends in public expenditure should be measured initially and private expenditure on agri-environmental goods, services and conservation added later. The OECD Producer and Consumer Subsidy Equivalents (PSE/CSE) database can be used to estimate public expenditure on agri-environmental measures (the government budget is generally the main data source). This indicator is broader than the related indicator presented in the Farm management section of this paper (Expenditure on agri-environmental research).

- *Interpretation:* "Sufficient" financial resources are necessary to enable farmers to adopt environmental farm management practices. When financed from private sources, it becomes an indicator of farmer awareness, priorities, etc.

- *Further refinement:* Estimation of private expenditure on environmental goods, services and conservation, to be developed in the longer term as no data are available at present. A definition is required to determine what constitutes "sufficient" financial resources.

Farm Financial Equilibrium

- *Definition:* The equilibrium between the net farm operating profit after tax (NOPAT, i.e. farm monetary receipts), and the cost of capital (i.e. financial costs to the farm).

- *Method of calculation:* Measured in current and real terms per farm.

The net operating surplus or net farm operating profit after tax (NOPAT, i.e. farm monetary receipts) can be calculated as:

+	Final agricultural output
−	Intermediate consumption
=	Gross value added (market prices)
+	Subsidies
−	Taxes linked to production
=	Gross value added (factor cost)
−	Depreciation
=	Net value added (factor cost)
−	Compensation of employees
=	NOPAT (Net operating surplus)

The cost of capital can be calculated as:

+	Interest payments
−	Tax credits
=	Cost of debt
+	Debt repayments
+	Drawings
−	Compensation for family labour
−	Off-farm income
=	Cost of capital

Where part or all of the property is leased, the rent (net of tax credits) would also be added to the cost of capital, since it is money withdrawn from the business to meet the requirements of the owners of resources used to generate the operating profit. The data sources for these calculations include the OECD Structural Indicators and Agricultural Accounts database.

- *Interpretation:* This is an indicator of "economic sustainability" and must be seen in conjunction with indicators for other agri-environmental areas, in order to assess the overall sustainability of the farm. The indicator shows whether the farmer has made adjustments to the operating profit, cost of capital or both, in order to maintain financial sustainability.

- The cash cost of equity can also be used to track social changes. This is to say the closer the cash cost of equity moves to zero, the more the family is treating the farm as a dwelling place, a place from which to generate both on and off-farm income, rather than providing a return on farm ownership apart from any capital gain. A negative trend in the cost of equity suggests an imminent structural change. This could occur, for example, when the next generation joins the business especially if the following generation has no wish to realise a return less than the value of their labour and management.

- *Further refinement:* Links between trends in this indicator and poverty levels in certain agricultural regions of OECD countries might be further explored.

Further development of indicators

Adjusted Farm Financial Equilibrium

- *Definition:* Adjusting farm financial resources for changes in natural resource depletion and pollution, for example, soil erosion and nutrient soil surface balance.

The indicator shows if the farm is maintaining its financial resources at the cost of resource depletion or pollution. This sustainable cost approach extends farm financial equilibrium to include resource management. A positive trend indicates that the farm is not only maintaining financial and environmental resources but also reinvesting for growth.

3.3   *Socio-cultural Issues (Rural Viability)*

Policy issues

"Socio-cultural issues" is an over-arching concept that includes, for example, demographics, rural traditions, norms and values, folkcustoms, social interactions, attitudes, community structure and work. In an agricultural context, the term specifically includes the impact of agriculture on rural communities as well as the contribution of rural life and culture to agricultural development. Many socio-cultural indicators do not have an irrefutable direct relation to sustainable agriculture. The objective here is to select indicators that can be tied most closely to sustainable agriculture, and which complement other agri-environmental indicators.

Indicators can help policy makers to understand the link between socio-cultural issues and sustainable agriculture, by addressing i) environmental sustainability, ii) economic viability, and iii) socio-cultural acceptability. The indicators below address the following aspects related to rural viability: agricultural income and the entry of new farmers (by age and gender) in the agricultural sector. Indicators on social organisation or social capital could be developed in the longer term.

Main points discussed

The Group considered "rural viability" to be a better term to describe this social dimension of sustainable agriculture than "socio-cultural issues". The Group noted that some OECD work on data and measurement is already underway related to socio-cultural issues, although the focus in general is on rural development and structural adjustment in agriculture, rather than sustainable agriculture per se.[5]

Further conceptual work may be required to examine the linkages between the socio-cultural, economic and environmental components in the Driving Force-State-Response framework and also to consider the spatial (disaggregated) aspects of these links in developing appropriate indicators.

---

5.   See, for example, OECD (1996), *Territorial Indicators of Employment — Focusing on Rural Development* and OECD (1994), *Creating Rural Indicators for Shaping Territorial Policy,* Paris.

Several indicators linking socio-cultural issues to rural development were discussed, which may form contextual indicators or sets of basic data, including the following:

- changes in agricultural land use, discussed under the report on biodiversity, wildlife habitat and landscape;

- changes in the number of full-time farmers: time series showing changes of farm employment in rural societies as agriculture usually plays a key role in the viability of rural communities.

Recommendations

Indicators for the short term

Agricultural Income

- *Definition:* Share of agricultural income in relation to total income of rural households.

- *Method of calculation:* The ratio can be calculated as:

$$\frac{\text{median agricultural household income}}{\text{median rural household income}} * 100$$

This indicator is related to the indicators on farm financial resources and the ratio shows the degree of integration of agricultural production and farm incomes in the rural economy.

Possible data sources to measure include the OECD's: Structural Indicators, Agricultural Accounts and Rural Development database.

- *Interpretation:* This indicator illustrates changes in rural livelihood. If agricultural household incomes are significantly below rural incomes, entering the agricultural sector will no longer be attractive. On the other hand, agricultural incomes may be higher than rural incomes as a result of agricultural policy transfers. Agricultural incomes may also increase in line with rural household incomes, thus further complicating the interpretation.

- *Further refinement:* The definition of agricultural and rural households vary across OECD countries and data on rural incomes, in particular, are not readily available. Consequently, it might be better to measure the share of on-farm income in relation to total farm income. This indicator of pluriactivity would also have a more straightforward interpretation thus removing some of the ambiguity.

Entry of New Farmers into Agriculture

- *Definition:* Number of farmers, according to age and gender, entering the agricultural sector.

- *Method of calculation:* Demographic pyramids showing new farmers by gender and age at 5-10 year intervals. Calculated at 5-year intervals, drawing on data from national authorities.

- *Interpretation:* This indicator is used as a proxy for the "attractiveness" of career opportunities within farming for young people. The rationale for this indicator is that any profession that does not appeal to the young could become unsustainable in the long term.

- *Further refinement:* Ideally, the attitudes of farmers and the general public to farming could complement this indicator. This might be developed through farmer interviews on succession, general opinion polls, etc.

Further development of indicators

Social Capital in Agricultural and Rural Communities

- *Definition:* The strength of social institutions and formal/informal networks, voluntary organisations, etc., in agricultural and rural communities.

  There is an extensive and evolving interest in addressing the various aspects of social capital and their importance, as a key component in sustainable agriculture (see the paper by David Pearce in this report). An increasing trend in social capital could mean that:

  - societies and institutions are getting more cohesive and transaction costs are being reduced;

  - trust in society is increasing and individuals can reduce costs in "protecting" themselves; and

  - there is less need for policing and monitoring, etc.

In the context of sustainable agriculture, social capital includes the preservation of rural communities and the rural "way of life". This, however, does not necessarily mean that all aspects of modern day agriculture should be included in social capital. Fundamental work to define the concept of social capital is still needed, before relevant indicators can be developed.

PART III:

AGRI-ENVIRONMENTAL INDICATORS AS A POLICY TOOL

CROSS-CUTTING ISSUES IN DEVELOPING AGRI-ENVIRONMENTAL INDICATORS

by
Andrew Moxey
University of Newcastle-upon-Tyne, United Kingdom

Executive summary

Increasing concern over environmental issues is leading to pressure for policy reforms to protect and enhance the agri-environment. Reacting appropriately to this pressure requires information on agri-environmental conditions and their responsiveness to agricultural activities induced by different market or policy signals. Agri-environmental indicators (AEIs) are viewed as one means of collating and presenting such policy-relevant information.

Consensus on the choice and construction of AEIs has, however, yet to be achieved. This is due to two sets of issues. First, AEIs have to address interactions of both socio-economic and environmental factors. Consequently, the debate is inevitably complicated, not least because many agri-environmental goods and services are not valued within conventional markets. Second, concern over agri-environmental issues has arisen only relatively recently. Consequently, there has not yet been sufficient debate to produce a consensus over many of the issues involved.

The Driving Force-State-Response (DSR) framework adopted by the OECD seeks to encourage consensus by facilitating debate over AEIs within a common, formal structure. Development within this framework may be viewed as proceeding in three stages.

1. The identification and measurement of underlying agri-environmental linkages and conditions.

   This involves dialogue between environmental scientists, social scientists and policy makers to agree upon areas of concern, current understanding of causality and availability of data to describe conditions and linkages. Issues are encountered here with respect to the theoretical understanding of agri-environmental systems and the availability of data to describe and monitor them adequately. In particular, with respect to which variables should be measured, how, when and where.

2. The incorporation of physical AEIs into an economic framework to allow explicit consideration of trade-offs between agri-environmental conditions and productive capacity, be that within agriculture or indeed elsewhere in the economy.

   This involves consideration of methodologies for ranking and valuing agri-environmental goods and services. In particular, how can an array of physical, chemical, biological and socio-economic variables be summarised, or combined, to ease identification and analysis of (marginal) trade-offs.

3. The extension of AEIs to the policy making arena as a decision support tool for exploring the trade-offs involved in alternative policy scenarios.

   This involves consideration of how AEIs should be interpreted, and raises issues concerning the setting of threshold and target levels against which to judge an AEI value. It also raises issues regarding the criteria for adoption of particular indicators. It is suggested that adoption depends on: good underlying theory; good underlying data; and general political acceptance.

Adoption and interpretation of AEIs may depend as much upon transparency in their design and presentation as upon their theoretical and empirical validity. That is, openness in the design and presentation of AEIs is likely to encourage their acceptance and adoption by policy makers. This suggests that the public debate facilitated by the DSR framework is to be welcomed.

1. Introduction

Increasing concern over environmental issues is leading to pressure for policy reforms to protect and enhance the agri-environment. Reacting appropriately to this pressure requires information on agri-environmental conditions and their responsiveness to agricultural activities induced by different market or policy signals. Such information is, however, typically fragmented and not in a form immediately amenable to policy makers.

Agri-environmental indicators (AEIs) are viewed as one means of overcoming this problem by collating and presenting policy-relevant information. In so doing, it is hoped that AEIs will contribute to a better understanding of the linkages between agriculture and the environment, thereby assisting in the design, implementation, monitoring and evaluation of policies.

Bonnen (1989) suggests that indicators are part of a continuum ranging from raw data through to knowledge which encompasses validated information around which a broad consensus has formed. Indicators fall within these extremes and represent raw data that have been combined within a conceptual framework to offer a particular summary view of, in this case, the agri-environment.

As such, indicators are just one type of information that is available to policy makers. Their attraction lies in offering a formalised, routine information source rather than an ad-hoc information source. That is, once agreed, a standard methodology for their construction can be applied, and values compared with pre-specified threshold and target levels, to judge improvement or deterioration in the agri-environment.

However, whilst the development of AEIs may be a reasonable response to the information needs of policy makers, it is not without its own conceptual and empirical problems. The aim of this paper is to review briefly some developmental and usage issues which cut across individual AEIs. Not all of the issues are necessarily unique to AEIs, but nevertheless merit discussion here.

Specifically, this paper will consider issues relating to: data measurement; indicator aggregation and valuation; and interpretation and adoption. First, however, the next section discusses the definition of indicators and describes the importance of the common framework adopted for developing and reporting AEIs within OECD.

2.   Indicators and frameworks

2.1   Indicators

Definitions of indicators as a concept (let alone specific indicators), vary widely. Different agencies and different authors each supply their own working definitions. Hence, Gallopín (1997) reports that indicators are referred to variously as: variables; parameters; measures; statistical measures; proxy measures; values; metres or measuring instruments; fractions; indices; a piece of information; empirical models of reality; signs.

The common theme running through this list is that indicators are a vehicle for collating, summarising (or otherwise simplifying) and communicating information about something that is of importance to decision makers or stakeholders. This suggests that indicators may be distinguishable from other forms of information only by their relevance to, and acceptance by, decision makers and the stakeholders that they represent: information is elevated to the status of an indicator by its user(s).

This is an important point and suggests that choice of an indicator may involve public and political acceptability as well as scientific rigour. In contexts where public policy decisions have some historical precedence, consensus has typically emerged over choice of indicators. Thus, for example, standard measures of national inflation and economic activity are widely agreed and accepted. In the case of AEIs, however, such consensus has yet to emerge. This is for two main reasons:

- AEIs have to address interactions of both socio-economic and environmental factors. Consequently, the debate is inevitably complicated, not least because many agri-environmental goods and services are not valued within conventional markets.

- Concern over agri-environmental issues has arisen only relatively recently. Consequently, there has not yet been sufficient debate to produce a consensus.

Hence, as Linster (1997, p. 163) notes, "an important lag remains between the demand for environmental indicators, the related conceptual work and the actual capacity for mobilising and validating data."

To facilitate the process of arriving at a consensus on a set of preferred AEIs, a common framework for developing and presenting AEIs is required. Without such a framework, comparison and review of the many different candidate AEIs currently being developed would be very difficult. This is particularly true given the variation in environmental endowments and agricultural structures, and associated problems, both within and between countries. The OECD has adopted the Driving Force-State-Response (DSR) framework as a means of facilitating the AEI debate (OECD, 1991, 1997).

2.2 *Driving Force-State-Response (DSR) framework*

The DSR framework recognises explicitly that agri-environmental interactions and linkages are complex and multi-faceted, and provides a structure within which individual indicators can be placed in context. In particular, the framework identifies three types of potential AEIs:

- State AEIs. Biophysical conditions such as water purity, soil texture and biodiversity that describe the state of the agri-environment.

- Driving-Force AEIs. Natural and socio-economic forces which influence (drive) the state of the agri-environment. For example, meteorological conditions and population growth but also government policy and farm management practices.

- Response AEIs. Responses of governments, consumers and producers to observed agri-environmental states and driving forces. For example, producers may change input and output combinations, consumers may alter their purchasing patterns and government may change policy prescriptions.

Each of the three indicator types encompasses many different individual indicators. Moreover, as with any classification, the boundaries between drivers, states and responses may be unclear in some cases. For example, the vegetation cover on a parcel of land might be interpreted as a state variable which is influenced by drivers such as agricultural commodity prices, yet could also be interpreted as a driver itself which influences other state variables such as soil condition or biodiversity. In addition, some indicators may be combinations of other indicators. For example, wildlife habitats embody climate, soils, and vegetation.

The value, therefore, of the DSR lies not so much in the precise categorisation of individual indicators, but rather in the provision of a common framework within which indicators can be presented and debated. The use of such a framework helps to highlight conceptual and methodological issues encountered.

2.3 *Agri-environmental Indicator (AEI) development*

Within the DSR, AEIs are being developed as an information input into the policy decision-making process. At present, information on the agri-environment is fed through to policy makers in a relatively ad hoc manner. That is, information arrives from a variety of sources such as statistical survey data, mathematical models and expert opinions. AEIs are viewed as offering a more formal and routine manner of gathering and communicating information. That is, once agreed upon, a set of AEIs can be used consistently. Developing AEIs to fulfil this purpose may be viewed as a three-stage process (Figure 5).

Figure 5. Diagram showing three stages in agri-environmental indicator development

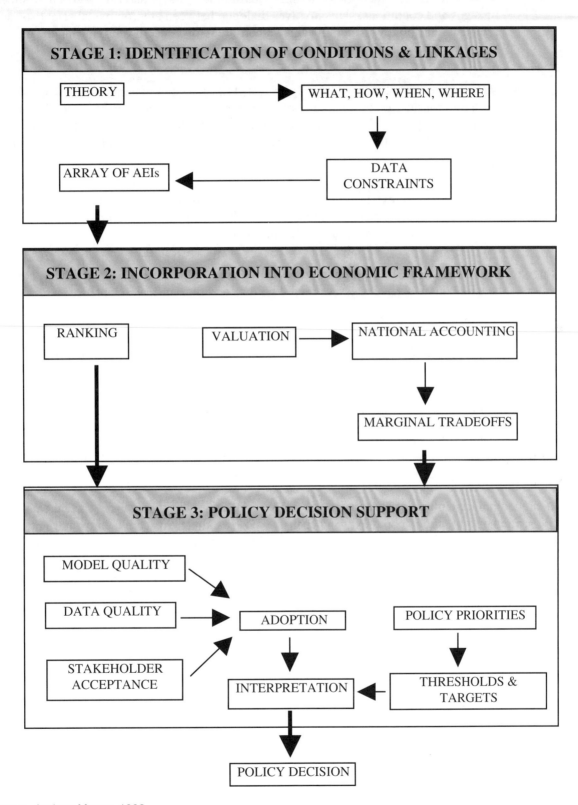

Source: Andrew Moxey, 1998.

- Stage one is the identification and measurement of underlying agri-environmental linkages and conditions. This involves dialogue between environmental scientists, social scientists and policy makers to agree upon areas of concern, current understanding of causality and availability of data to describe conditions and linkages.

- Stage two is the incorporation of physical AEIs into an economic framework to allow explicit consideration of trade-offs between agri-environmental conditions and productive capacity, be that within agriculture or indeed elsewhere in the economy. This involves consideration of methodologies for ranking and valuing agri-environmental goods and services.

- Stage three is the extension of this to the policy making arena as a decision support tool for exploring the trade-offs involved in alternative policy scenarios. This involves consideration of how AEIs should be interpreted.

All three of these stages in AEI development raise some issues which cut across individual candidate indicators. Stage one raises particular issues regarding data and their measurement.

3.   Data measurement

3.1   What

Identification and measurement of underlying agri-environmental linkages relies upon both theoretical understanding and data availability. Theoretical understanding is required to guide the choice of which variables to measure, and the manner in which they should be measured. Data arising from empirical measurement are required to actually implement construction of an indicator.

As theoretical understanding of agri-environmental systems continues to advance (e.g. Jakeman *et al.*, 1995), awareness of their complexity deepens. For example, research into nutrient cycles, hydrological cycles and commodity market interactions continues to inform research and policy communities alike as to the relative importance of, and jointness between, different components of agri-environmental systems. This assists designers of AEIs in selecting which physical, chemical, biological, and socio-economic variables should be focused upon when constructing AEIs.

3.2   How

In addition to identifying which variables to focus upon, theoretical understanding also guides how variables should be measured. That is, agri-environmental variables are vector rather than scalar measures, with several dimensions. In particular, consideration has to be given to quantity and quality aspects, plus inter-temporal and spatial dimensions.

Whilst it is possible to simply record the presence or absence of a variable, it is more common to quantify its magnitude. Hence, for instance, the presence of landscape features such as ponds and hedgerows is normally supported by a measure of their number or length. Similarly, usage of a pesticide is usually quantified rather than simply expressed as usage or non-usage.

Such simple quantification may not, however, convey important information regarding the condition of the variable. For example, ponds may be old or new, shallow or deep, stagnant or healthy. Hedges may be old or new, broken or complete, uniform or varied in species. Pesticide applications may be by spray or contact. Such qualitative differences can have significant agri-environmental impacts. An analogy may perhaps be drawn here with economic employment data where the type of job (e.g. full or part-time, permanent or temporary) is considered a useful piece of information to accompany the number of jobs reported.

3.3   When

Agri-environmental systems change over time, particularly in response to seasonal fluctuations. This means that measurement values at one time of year may not be directly comparable with, or as meaningful as, values taken at another time of year. For example, the water content of soil fluctuates over the year in response to changing rainfall levels and vegetation cover. Winter measurements are not comparable with summer measurements in terms of, for example, describing plant stress and the subsequent demand for irrigation.

Similarly, the nitrate leaching rate from a given application of fertilizer will depend not only upon the quantify applied, but also the timing of the application and the prevailing meteorological conditions: using annual application and rainfall data will give a different picture to using data temporally disaggregated to daily events.

Consideration also needs to be given to the time frame over which variables change. Some may change relatively slowly, for example soil type, whilst others may change relatively rapidly, for example soil moisture content. Care also needs to be exercised in distinguishing genuine change (a trend) in the system from simply stochastic variation. That is, an apparent change in a variable's value may actually be within its usual bounds of variation. For example, average temperatures can fluctuate significantly year-on-year without necessarily displaying long-term change.

3.4   Where

Finally, due to the spatially extensive nature of agricultural production and the spatial heterogeneity of agriculture and its underlying environmental resource base, agri-environmental variables have a spatial dimension. This means that they need to be expressed over an appropriate spatial unit of measurement. This may be a hectare, a catchment, an agro-ecological zone, a trading block or any other unit, depending on the variable of interest.

For example, livestock per hectare is an indicator of grazing intensity and total fertilizer usage within a catchment is a indicator of nitrate loading to the sea. Choice of spatial unit is made more complicated by the fact that agri-environmental systems display spatial connectivity. That is due to, for example, hydrological systems, the dispersal of plants and animals or the movement of farm produce, and what happens at one location may not be independent of what happens at another location.

Theoretical understanding of agri-environmental systems thus offers some guidance as to which variables are important, and how they should be measured and at which points in time and space. Choice of variable(s) and measurement details for inclusion in an indicator is, however, not determined solely by theoretical considerations. Rather, choice is constrained by data availability.

3.5   Data constraints

Historically, agri-environmental issues have not been considered sufficiently important to justify separate data collection and relatively few AEIs are based on specially commissioned data collection exercises. Consequently, although this situation is now changing (e.g. Stanners and Bordeau, 1995), most existing datasets do not relate specifically to agri-environmental issues. Hence, theoretical data preferences may be compromised to some extent by practical, empirical restrictions.

Nevertheless, although data relating specifically to agri-environmental linkages may be missing or incomplete, data collected for other purposes are, by comparison, readily available. For example, whilst direct measurements of biodiversity, pollutant contributions and landscape features may be scarce, data relating to, for instance, agricultural land covers, are relatively abundant. This means that the scope for AEI design is still good, provided that some consideration is given to the quality of secondary data and to the manner in which they are used.

In particular, especially given the need for international comparisons of AEIs, reliance on secondary data raises two issues:

- Data from different sources often have slightly different definitions, even when supposedly describing the same thing. For example, definitions of "rough grazing" and "pasture" can vary between countries according to what is the best and worst grassland found it that country. Consequently, compiling consistent data coverage can be difficult.

- Different data items from different data sources are often reported for varying spatial and/or temporal sampling frames and scales, making their collation and combination difficult. For example, data on nitrogen fertilizer usage may be reported annually, yet hydrological data may be reported daily.

The problems raised by these issues are difficult, but not impossible to deal with. Indeed, advances in information technology, particularly in the fields of relational database management systems (RDBMS) and geographical information systems (GIS) are facilitating pragmatic utilisation of many digital datasets (Jakeman *et al.*, 1995).

For example, the availability of remote sensed land cover data provides a detailed snapshot of agricultural land cover which can be combined relatively easily with data on soil types, rainfall patterns and typical farm management practices to estimate the risks of nitrate leaching or soil erosion over a given area (Allanson *et al.*, 1993). Similarly, GIS can be used to identify landscape patterns and integrity from land cover data and simple ecological rules (Pastor and Johnston, 1992; Jones *et al.*, 1997).

The robustness of agri-environmental information compiled from secondary data rests on both the quality of the secondary data sources and any assumptions made in their usage. For example, combining secondary data reported at different spatial scales can incur various methodological problems (Robinson, 1950; Oppenshaw, 1984).

In addition, AEIs inevitably embody some uncertainty. That is, imperfections in the theoretical understanding of the agri-environmental systems and errors in data measurement both contribute to imprecision in indicators. Provided that the designers and users of AEIs are aware of such problems, and report how they were addressed (e.g. Wallace, 1994), there is no reason why, however, a full range of AEIs should not be developed.

Perhaps more problematically, due to the complexity of agri-environmental systems, users of AEIs will be faced with a bewildering array of physical, chemical, biological and socio-economic indicators. Hence, once a full set of AEIs has been developed, the next stage is to incorporate them into an economic framework. This entails consideration of how they may be aggregated together to ease subsequent interpretation.

4.   Aggregation and economics

4.1   Ranking and weighting

As a primary industry, agriculture generates crop and livestock commodities through the application of various inputs to the land base resource. This process influences the land resource, changing not only its appearance, by virtue of shifting from natural to semi-natural or alien cover types, but also less visible characteristics such as nutrient balance, soil texture and water content. These changes have physical, chemical and biological impacts on the agri-environment, some of which are regarded as positive, others as negative.

For example, changes in types and spatial juxtaposition (or mosaic) of land covers can alter landscapes and fragment wildlife habitats, whilst associated changes in land management practices can lead to air and water pollution. The purpose of AEIs is to assist policy makers by clarifying trade-offs between these different aspects of the agri-environmental system. For example, less intensive crop production may improve the agri-environment but produce less food and reduce farm incomes.

The complex, multi-functional nature of agri-environmental systems does, however, inevitably throw up an array of candidate AEIs competing for policy maker's attention. This suggests the need for procedures for summarising indicators themselves in order to clarify trade-offs. One approach is to attach rank order to competing indicators, to identify the most important. This is inevitably a subjective process, and, guided by theoretical understanding of local agri-environmental systems, will vary across countries. For example, soil erosion will be more of a priority in Canada than in Europe.

Alternatively, separate indicators can be combined through a weighting scheme to generate a composite indicator, perhaps in the form of an index. Here, problems are encountered due to different units of measurement across different indicators. For example, land use measured in hectares, soil depth in centimetres, and pesticide usage in kilograms of active ingredients.

One approach is to score each indicator value according to some categorical scale, and then simply add up the scores obtained. For example, the United States Department of Agriculture Environmental Benefits Index (USDA, 1997) assigns scores to seven different factors to generate a composite value for decisions regarding enrolment of land into a conservation scheme.

Alternatively, if some common unit of measurement can be found, the separate indicators can be combined more directly. This common unit might be physical, for example, weight or volume, but is more likely to be economic (Atkinson, 1995). That is, valuing all items in a common monetary currency unit allows aggregation across many millions of individual activities and items.

4.2   Valuation

Expressing all AEIs in monetary terms would clarify policy trade-offs. Unfortunately, whilst production commodities are valued through market transactions, most agri-environmental effects are not. Consequently, surrogate valuation techniques have to be employed. For some AEIs, valuation can be based on lost productive capacity or the cost of remedial action. For example, soil erosion can be valued through estimation of reduced yields and subsequent revenue, water purification costs can be used to estimate the value of nitrate pollution.

For other effects, values are largely non-use or preservation values and techniques such as Contingent Valuation Methods have to be deployed (Hutchinson, 1997). Such techniques have been widely used and endorsed by leading economists and, although imperfect, represent perhaps the best way of explicitly quantifying trade-offs to assist policy decisions.

In particular, without converting AEIs to a common unit of currency, it is impossible to engage in marginal (economic) analysis of trade-offs. That is, physical, chemical and biological AEIs are expressed in terms of averages rather than marginal conditions. For example, the average biodiversity in a region or the average emission of greenhouse gases per hectare, not the additional biodiversity gained (or lost) in a region or the additional emission of greenhouse gases from the last hectare entered into production. This is a serious handicap for policy makers attempting to identify trade-offs.

4.3   National (economic) accounting

The preceding section on economic valuation of agri-environmental costs and benefits arising from agricultural activities leads onto a wider consideration of the economic role of agriculture. Whilst internal debates may be concerned with trade-offs and interactions between agriculture and the agri-environment, it is important to place these in the context of the broader economy. That is, agriculture interacts with other economic sectors.

This means that AEIs may need to be extended to encompass other sectors and/or indicators developed for other sectors may need to be considered alongside AEIs: environmental issues cut across administrative and sectoral classification boundaries. To some extent, this fact has already been acknowledged in the parallel construction of indicators for sustainable development. For example, in the work of the UN's Commission on Sustainable Development (UNCSD, 1996) or the UK's Department of the Environment (Department of the Environment, 1996).

Most obviously, agriculture is intricately linked with, for example, both upstream agricultural input industries and downstream food processing industries. In addition, however, agriculture competes for resources (most obviously land) with other, unrelated sectors such as transport, energy and housing. Moreover, despite not being agricultural, these other sectors may affect the state of the agri-environment. For example energy, transport and housing contribute to air and water pollution.

Consideration of such trade-offs between agri-environmental (public good) benefits and commodity (market) benefits thus can not be confined solely to the agricultural sector. Rather trade-offs need to be recognised with other sectors. For example, more housing to meet residential demand, or improved roads networks, deliver benefits which need to be weighed against the loss of agricultural productive capacity and possible agri-environmental damage. This suggests that agri-environmental goods and services should be included in national accounting frameworks.

Historically, national accounting has been restricted to measurement of marketed commodities, with the resulting measure of economic activity, and its distribution across different sectors, being used as an input to government policy deliberations. This approach has been criticised, however, for excluding areas of economic activity and welfare falling outside of the normal market system.

The value of domestic services of all kinds were an early noted omission, and more recently the debate has moved on to wider externalities, including agri-environmental items. These are of particular interest in the so-called "greening" of the accounts to accommodate both current and future environmental goods and services (Whitby and Adger, 1996, 1997; Atkinson, 1994, 1995). Some success has been achieved at including natural resource depletion into national accounts, but items such as landscape and biodiversity have yet to be addressed adequately.

Costanza *et al.* (1997) present possibly the most ambitious attempt to place environmental goods and service within an economic accounting framework, valuing ecosystem services at the world level, concluding that they are worth a total of US$33 trillion in comparison with a world Gross National Product of US$18 trillion. Whilst the precise numbers may be debatable, such exercises do serve to highlight the possible relativities between the value of conventionally measured economic activities and previously ignored externalities.

Incorporating AEIs into a national accounting framework would, therefore, clarify some of the trade-offs between agri-environmental conditions and costs or benefits elsewhere in the economy. The weighting system used, whilst not beyond criticism, is at least transparent and can be amended by anyone with better information.

The weakness of such calculations will remain the insecure basis of the valuations they embody. The difficulties to be overcome are thus in agreeing upon valuation techniques or, if such valuations are to be avoided, on how to compare physical AEI measures with economic activities measured in monetary units, and on how to conduct marginal analysis.

5.   Interpretation and adoption

5.1   Thresholds and targets

The third stage in AEI development is their inclusion in the decision-making process. This raises issues of interpretation and adoption. On its own, an indicator value is of little use. Rather, it has to be compared with some pre-specified value. That is, certain values of an indicator assume special significance. Such values may be referred to as: thresholds; targets; and benchmarks or reference levels (Gallopín, 1997). These terms are not necessarily interchangeable.

Thresholds are scientifically determined boundaries at which a significant change occurs to a system, for example, the pollutant concentration above which human health will be affected, or the population density below which a species becomes non-viable. Thresholds are particularly important in an agri-environmental context given the propensity of ecological systems to "flip" from one state to another.

That is, once an agri-environmental system attribute goes beyond a certain minimum or maximum level, the state of the system may change dramatically. For example, soil erosion may be negligible provided that a given minimum proportion of land is covered by vegetation, but suddenly becomes a problem if vegetation cover shrinks below this minimum.

Targets allude explicitly to managerial intentions. That is, they are set in the context of a decision-making process and are expected to be achievable through some reasonable course of action. For example, stabilisation of greenhouse gas emissions to 1990 levels by the year 2000. The target set will vary according to several factors, including the current indicator level and scientific threshold levels, but also political or societal consensus. This last point is important.

In order to operationalise agri-environmental policies, policy makers need some notion of the dividing line between environmental benefits and environmental harm. Without such a benchmark or reference level, it is impossible to determine whether resource users should be rewarded or penalised because it is unclear whether they are providing benefits or causing damage.

For example, conservationists may view the draining of a wetland as causing harm meriting some penalty, whilst the owner of the land may view not draining the wetland as providing a benefit that merits reward (after Bromley, 1997). That is, different groups within society typically hold different views on what constitutes an environmental benefit or harm. Views held by different groups may vary spatially, especially across international boundaries where cultural differences are evident.

Hence, as Bromley (1997) argues, reference levels for agri-environmental conditions are a social construct, depending very much upon current conditions and the distribution of property rights within the agri-environmental system. That is, reference levels, and therefore targets, for AEIs may be determined as much by social and political factors as by objective scientific criteria. Again, this highlights the value of the DSR framework as a focal point for AEI development and identification of policy priorities.

5.2   *Agri-environmental indicator adoption and usage*

To identify a policy role for AEIs it is necessary to consider the way in which public decisions about the environment are taken. Confronted with the need to make a decision, Public Decision Makers (PDMs) will seek information from which they can assess possible outcomes. The more senior they are in the hierarchy the more political information they will require, the lower they are the more they will depend on technical and factual information. As technical information moves upwards through the hierarchy it will become more refined, analysed and reduced in volume (Hutchinson, 1997).

The attraction of AEIs is thus that they offer a formalised, routine method for collating and communicating policy-relevant information. That is, once an indicator (or set of indicators) has been adopted, a standard methodology can be employed on a regular basis to generate indicator values. These can then be compared with pre-specified threshold or target levels to judge whether some policy response is necessary.

Such interpretation, however, needs to be carefully considered. Indicators are imperfect representations of reality. They represent compromise designs, involving trade-offs between ease and cost of data measurability, scientific validity, transparency and relevance to users: at best they may be "optimally inaccurate". As such, no individual indicator can be universally applicable and may be open to abuse if used out of context (see Box 6 for an example of interpretation issues for a land use AEI).

This does not mean that AEIs are of no value. Rather, their value depends partly upon how carefully they are interpreted and the manner in which they are combined with other policy information. This means that policy makers need to be aware of the issues involved in AEI design and construction.

That is, in order to avoid misuse or abuse of indicators, it is essential that policy makers recognise assumptions, limitations and errors embodied in particular indicators. This suggests that careful consideration of adoption criteria is required.

Costanza *et al.* (1992) suggest that the policy relevance and likely adoption of information rest on three criteria:

- Quality of the underlying theoretical model.

- Quality of the underlying data.

- Degree of acceptance (by academics, practitioners, and stakeholders).

Hence, adoption of an AEI for policy purposes may require that not only is it theoretically sound and backed by reasonable data, but it should also enjoy popular support. That is, political acceptance is important in determining the choice of indicator: if indicators are to have a role in problem identification, problem acknowledgement, awareness raising, policy formulation, policy implementation, justification and evaluation, then they need to be accepted by decision makers and the public (MacNaughton *et al.*, 1997).

Acceptance criteria will be influenced by the manner in which AEIs are presented. In particular, two presentation characteristics are desirable:

1. *Transparency*. To be accepted by decision makers and stakeholders in general (see Chapters 8 and 40 of Agenda 21, UN, 1993), both the reasoning behind the choice of an indicator and the process by which it is derived from available data need to be transparent. This places the onus upon designers of indicators to be not only competent scientifically, but also to be able to argue or present indicators to a (possibly) non-technical audience. This poses difficulties where the underlying science may be complex. It also poses difficulties where the degree of processing of underlying data is significant.

2. *Relevance and ownership*. To be used as an indicator, a piece of information has to be relevant to a decision maker. Moreover, the decision maker has to perceive some self-capability to monitor the indicator and/or influence the value of the indicator through (positive) action. If an offered indicator does not appear relevant to a decision maker, or is perceived to be beyond the decision maker's sphere of influence, it is unlikely to be accepted.

   These issues point to the need for dialogue between the designers of indicators, users of indicators, and other stakeholders. Designers need some appreciation of the decision (e.g. policy) making process and the managerial opportunities and constraints, including the fact that elevation of information to the status of an indicator will possibly induce behavioural responses from people, i.e. reactivity (Fitz-Gibbon, 1990). Conversely, decision makers need some appreciation of the theoretical and empirical basis for the choice and design of indicators. In this respect, the DSR framework serves as a useful focal point for debate.

Box 6. Land use as an example agri-environmental indicator

Agricultural land use data may be regarded as a potential agri-environmental indicator. That is, agriculture is a spatially extensive economic activity, impacting upon the environment through both the land cover(s) that it imposes and the manner in which land cover is managed.

For example, arable cropping alters soil conditions and landscape appearance through the imposition of a non-natural vegetation cover and the usage of agri-chemicals such as nitrogen fertilizers. These impacts may be negative or positive. For example, fertilizer usage may lead to nitrate pollution, but the mosaic of crop covers may provide an attractive landscape. Monitoring land use may thus provide a reasonable measure of various aspects of the agri-environment.

Measuring agricultural land use typically entails combining secondary data from various sources. For example, land cover data are generally readily available in many countries, either from regular censuses of agriculture or, more recently, from satellite imagery. These data do not, however, reveal land management. Rather management information such as fertilizer usage rates, crop varieties and field boundaries have to be taken from farm or ecological survey data.

Such data combinations may not be straightforward due to differences in sampling frames and data definitions. For example, land cover data may span a range of environmental conditions (e.g. soil types) which are not reflected in the sampling frame for farm management data which focuses on, for instance, farm type and size. Similarly, ecological definitions of, for example, grassland tend to be narrower than the commonly employed farming definitions. Combining secondary data therefore requires some assumptions to be made regarding how best to reconcile data differences.

Interpretation of land use indicator values also requires consideration of the underlying theoretical models and data: what does the indicator actually represent. Thus, for example, the area of farm land enrolled in conservation schemes could be taken as an indicator of environmental protection. This would, however, neglect the possibility that land enrolled was not necessarily managed in an appropriate manner. A better indicator would also include some measure of farmer compliance with management prescriptions.

Land use indicator values also need to be set against threshold and target levels. For example, a decrease in the area of agricultural land area due to urban development may be regarded as a good thing or a bad thing, depending on whether priority is given to retaining agricultural land or providing residential housing.

In some cases, priorities may be difficult to identify. For example, low lying agricultural land reverting to saltmarsh may be viewed as a bad thing to be avoided through the maintenance of sea defences, or it may be viewed as a good thing to be encouraged through set-aside schemes: a disharmony that currently exists between different policies in the UK.

Such ambiguous interpretations highlight the importance of clear statements of policy priorities, ideally represented as thresholds and targets. In addition, they demonstrate the problem of attempting to compare indicators for different policy areas. That is, even for the previous example of low lying agricultural land, comparison of the two outcomes (e.g. loss of productive agricultural capacity vs. habitat improvement) is difficult. Identification of the trade-off really needs both outcomes to be valued in monetary terms.

6. Conclusions

Increasing concern over environmental issues is leading to pressure for policy reforms to protect and enhance the agri-environment. Reacting appropriately to this pressure requires information on agri-environmental conditions and their responsiveness to agricultural activities induced by different market or policy signals. Agri-environmental indicators are viewed as one means of collating and presenting such policy-relevant information.

Consensus on the choice and construction of AEIs has, however, yet to be achieved. The Driving Force-State-Response framework adopted by the OECD seeks to encourage consensus by facilitating debate over AEIs within a common, formal structure. Development within this framework may be viewed as proceeding in three stages:

1. The identification and measurement of underlying agri-environmental linkages and conditions.

2. The incorporation of physical AEIs into an economic framework to allow explicit consideration of trade-offs between agri-environmental conditions and productive capacity, be that within agriculture or indeed elsewhere in the economy.

3. The extension of AEIs to the policy making arena as a decision support tool for exploring the trade-offs involved in alternative policy scenarios.

Each of these stages raise issues concerning the design and usage of AEIs. In particular, consideration needs to be given to data availability, aggregation of separate indicators, and interpretation.

Whilst some of these issues are not unique to agri-environmental problems, they still need to be addressed. Adoption of an open forum, encouraged by the DSR framework, assists in raising awareness of the issues and the manner in which they are dealt with.

This openness is a positive move. Policy decisions have to be made regarding agri-environmental problems. They can be made behind closed doors, or they can be made more openly. Given current national and international scrutiny of policies, for example under trade negotiations, openness is perhaps more desirable, if not inevitable.

This will, however, involve more public debate for, as Hardin (1968) noted: "It is when the hidden decisions are made explicit that the arguments begin." It is hoped that, by highlighting some cross-cutting issues, this paper has revealed some of the decisions hidden within AEIs.

BIBLIOGRAPHY

ALLANSON, P., A.P. MOXEY and B. WHITE (1993), "Measuring Agricultural Non-Point Pollution for River Catchment Planning", *Journal of Environmental Management*, Vol. 38, pp. 219-232.

ATKINSON, G. (1994), *Towards "Nature Conservation" Extensions to National Accounts: Some Possible Directions: a Report for English Nature*, Centre for Social and Economic Research on the Global Environment (CSERGE) University College, London and University of East Anglia.

ATKINSON, G. (1995), *Measuring Sustainable Economic Welfare: a Critique of the UK ISEW*, Centre for Social and Economic Research on the Global Environment, University College, London, Working Paper GEC 95-08.

BONNEN, J.T. (1989), "On the Role of Data and Measurement in Agricultural Economics Research", *Journal of Agricultural Economics Research*, Vol. 41, pp. 2-5.

BROMLEY, D. (1997), "Environmental Benefits of Agriculture: Concepts", in OECD, *Environmental Benefits from Agriculture: Issues and Policies — The Helsinki Seminar*, Paris.

COSTANZA, R., S. FUNCTOWICZ and J. RAVETZ (1992), "Assessing and Communicating Data Quality in Policy-Relevant Research", *Environmental Management*, Vol. 16, pp. 121-131.

COSTANZA, R. *et al.* (1997), "The Value of the World's Ecosystem Services and Natural Capital", *Nature*, Vol. 387, 15 May, pp. 253-260.

DEPARTMENT OF THE ENVIRONMENT, UNITED KINGDOM (1996), *Indicators of Sustainable Development for the United Kingdom*, HMSO, London.

FITZ-GIBBON, C. (1990), *Performance indicators,* Bera Dialogues No. 2, Multilingual Matters Ltd., Clevedon, Philadelphia, United States.

GALLOPÍN, G. (1997), "Indicators and their Use: Information for Decision Making", in B. Moldan and S. Billharz, (eds), *Sustainability Indicators,* Report on the Project on Indicators of Sustainable Development, John Wiley and Sons, Chichester, United Kingdom.

HARDIN, G. (1968), "The Tragedy of the Commons", *Science*, Vol. 162, pp. 1 243-1 248.

HUTCHINSON, W.G. (1997), "Environmental Benefits of Agriculture: Evaluation Methods to Measure and Monitor Change", in OECD, *Environmental Benefits from Agriculture: Issues and Policies — The Helsinki Seminar*, Paris.

JAKEMAN, A.J., M.B. BECK and M.J. McALEER (eds) (1995), *Modelling Change in Environmental Systems,* John Wiley and Sons, Chichester, United Kingdom.

JONES, B., J. WALKER, K. RIITTERS, J. WICKHAM. and C. NICOLL (1997), "Indicators of Landscape Integrity", in J. Walker and D. Reuter (eds), *Indicators of Catchment Health: a Technical Perspective*, CSIRO Publishing, Victoria, Australia.

LINSTER, M. (1997), "OECD Environmental Indicators", in B. Moldan and S. Billharz (eds), *Sustainability Indicators*, Report on the Project on Indicators of Sustainable Development, John Wiley and Sons, Chichester, United Kingdom.

MacNAUGHTON, P., R. GROVE-WHITE, M. JACOBS and B. WYNEE (1997), "Sustainability and Indicators", Chapter Eight, in P. McDonah and A. Prothero (eds), *Green Management: a Reader,* pp. 148-153, Dryden Press, London.

OECD (1991), *Environmental Indicators: a Preliminary Set*, Paris.

OECD (1997), *Environmental Indicators for Agriculture*, Paris.

OPPENSHAW, S. (1984), "The Modifiable Areal Unit Problem, Concepts And Techniques", in *Modern Geography,* Vol. 38, Geo Books, Norwich, United Kingdom.

PASTOR, J. and C. JOHNSTON, (1992), "Using Simulation Models and Geographic Information Systems to Integrate Ecosystem and Landscape Ecology", in R.J. Naiman (ed.), *Watershed Management: Balancing Sustainability and Environmental Change*, Springer-Verlag, New York, United States.

ROBINSON, W.S. (1950), "Ecological Correlations and the Behaviour of Individuals", *American Sociological Review*, Vol. 15, pp. 351-357.

STANNERS, D. and P. BOURDEAU (1995), *Europe's Environment: the Dobris Assessment* (with Statistical Assessment), European Environment Agency, Copenhagen.

UN (1993), *Programme of Action for Sustainable Development*, Reprint of the Agenda 21 Report of the United Nations Conference on Environment and Development, Rio de Janeiro, Vol. I, Resolutions Adopted by the Conference, United Nations, New York, United States.

UN COMMISSION ON SUSTAINABLE DEVELOPMENT [UNCSD] (1996), *Indicators of Sustainable Development: Framework and Methodologies*, United Nations, New York, United States.

USDA (1997), *Environmental Benefits Index*, October, Fact Sheet, Farm Service Agency, United States Department of Agriculture, Washington, D.C.

WALLACE, W.A. (ed.) (1994), *Ethics in Modelling*, Pergamon Press, Oxford, United Kingdom.

WHITBY, M.C. and N.W. ADGER (1996), "Natural and Reproducible Capital and the Sustainability of the Land Use Sector in the UK", *Journal of Agricultural Economics*, Vol. 47, No. 1, pp. 50-65.

WHITBY, M.C. and N.W. ADGER (1997), "Natural and Reproducible Capital and the Sustainability of Land Use in the UK: a Reply", *Journal of Agricultural Economics*, Vol. 48, No. 3, pp. 452-456.

USING AGRI-ENVIRONMENTAL INDICATORS TO ASSESS ENVIRONMENTAL PERFORMANCE

by
Paul J. Thomassin
McGill University, Quebec, Canada

Executive summary

Many governments are in the process of assessing their policies in terms of their sustainability. This creates a need for information in order to be able to assess policies in terms of their social, economic and environmental impacts. As a result, agri-environmental indicators (AEIs) are being developed to provide information on the environmental impact of government policy.

Modelling frameworks are being developed to integrate economic and bio-physical models in order to estimate the trade-offs between economic development and the environment. The models come in a variety of levels of aggregation: macroeconomic, sub-national, and ecological, all of which are required since the informational requirements for policy assessment are diverse.

The frameworks that are being developed have varying levels of detail and complexity. Generic frameworks, such as the Policy Evaluation Matrix (PEM), can be augmented to provide general estimates of the environmental impacts of policy. More detailed frameworks, such as the Economic-Ecological Input-Output Model, the biophysical Canadian Regional Agricultural Model (CRAM) or the Resource and Agricultural Policy Systems (RAPS), require substantially more data but will provide more detailed results. Other more qualitative frameworks can also be applied to assess policy impacts and, while these will not provide the detail of the other modelling frameworks, they can be undertaken with substantially less cost.

The informational requirements to assess policies in terms of their sustainability will be substantial. Establishing a set of standard AEIs would facilitate this evaluation and developing modelling frameworks that use AEIs will improve the accuracy of the estimated environmental impacts. A standard set of AEIs would also allow international comparison of environmental assessments. Model developments will have to occur at a variety of scales and levels of detail to supply the appropriate information to decision makers. Models that incorporate AEIs should be viewed as decision support systems that assist decision makers and the various stakeholders in the decision-making process.

In the short term, research is needed on data development, the modelling of feedback mechanisms between the economic and the bio-physical models, and the endogenising of environmental aspects of policies. In the long term, research will be needed on developing links between model output and decision makers in order that decision makers can pro-actively address environmental problems.

1. Introduction

Agri-environmental indicators (AEIs) are being developed as a means of integrating economic and environmental models to assist in the public policy decision-making process. Decision makers have traditionally analysed the economic impacts of programmes and policies, but recently environmental concerns are becoming increasingly important. As a result, decision makers want to know the answers to such questions as those listed below:

- What is the environmental impact of reducing subsidies to the agriculture sector?

- What are the environmental impacts of alternative agricultural policy instruments, such as direct payments versus market price support?

- What are the environmental effects of new policy initiatives?

- What are the environmental impacts of extending current policies into the future?

- What are the economic implications for the agriculture sector of meeting environmental targets, such as those set out in international agreements?

The answers to these types of questions will be used by decision makers to assess whether or not government policies satisfy the sustainable development objectives of society. Sustainable development requires policies to meet social, economic and environmental goals. The information that is needed for this assessment will have to have time, space, quantity, quality and equity dimensions associated with it.

Agricultural policy impacts occur in both the short and long term and must be accounted for. In a similar way, the spatial aspects of the analysis, either in terms of political boundaries or environmental considerations, must be appropriate. Information on the quantity and quality of the environmental impact must also be estimated. These estimates will have to take into account the complex biological and physical relationships as well as the human behavioural impacts. All of this information will have to be evaluated in the light of the social goals of government, and the economic viability of the sector.

AEIs can be used to generate information concerning the environment that can be incorporated into the decision-making process. Their development will increase our understanding of environmental processes and help in providing estimates of future policy impacts. This will assist decision makers and the various stakeholders with evaluating the trade-offs between environmental objectives, social goals and economic growth.

2. Using indicators for policy analysis

The information needed to assess the sustainability of agricultural policy needs to be based on the correct specification of the linkages between the agriculture sector and the environment. A framework that has been used for defining these linkages is the Driving Force-State-Response (DSR) framework (OECD, 1997a). This framework attempts to identify and quantify the complex interrelationships in these two areas. The first linkage is the interaction between agriculture practices and the bio-physical environment, while the second is between the human population, in particular farmers, consumers and policy makers and changing agricultural and environmental conditions.

The OECD (1997a) has identified a group of key agri-environmental indicators in order to quantify the relationships in the DSR framework, in particular, the linkages between agriculture and the environment. These indicators can be used to provide decision makers with information on the environmental consequences of actions taken as a result of changes in policy. In a similar fashion, the impact on the agriculture sector can be estimated if the AEIs are used to define the policy objective. This enables decision makers to identify the trade-offs between agricultural development and the environment and to predict the consequences of policy selection. It also allows policy analysts to monitor and project future environmental consequences of policy decisions.

Information required to assess the environmental impact of government policy will depend upon the situation being analysed. For example, the informational requirements will change with the institutional level, whether international, national or sub-national; as well as the types of policies being analysed, that is, generic or specific. The remaining sections of the paper will address these issues.

3.   Generic policy response analysis

Gardner (1987) developed a modelling framework to assess the impact of policy on the agriculture sector. The model estimates a number of relevant policy variables that can be used to determine the impact of policy changes. These policy variables can be analysed using a Policy Evaluation Matrix (PEM). The PEM provides a framework that estimates the impact of changes in policy on the level of support (as measured by the Producer and Consumer Subsidy Equivalents — PSE/CSE), on farm incomes, farm employment and transfer efficiency. PEM also estimates the distribution of costs and benefits between consumers, producers and between domestic and export markets, and changes in the quantity and value of production and trade. Table 5 provides an example of the information generated by the PEM framework.

Table 5. Information generated by the Policy Evaluation Matrix

Indicator	Description
1. Support	Estimates of PSE and CSE
2. Farm Income Source	Estimates of farm subsidies and market returns
3. Cost and Benefits	Estimates for taxpayers, farm households, input suppliers, domestic consumers and foreign consumers
4. Production and Trade	Estimates of the value and quantity of production, consumption and trade of relevant crops

Source: Author.

A pilot study is being undertaken by the OECD to test the feasibility of using PEM to evaluate the potential impact of changes in a number of agricultural policies. The countries included in the pilot study are currently: Canada, the European Union, Mexico and the United States.

Currently, the PEM framework used by the OECD has not been designed to evaluate the environmental impact of policy changes. This framework could be modified to incorporate AEIs into the policy matrix. One potential means of incorporating AEIs into the PEM is with a *gravity model*. A gravity model would allocate the changes in production to various regions within a country. Each of the regions would have average AEIs which would take into account the bio-physical properties and farm management practices of the region. An example of how the gravity model would work is as follows. If the PEM estimated that a change in policy would increase soybean production in Canada, then the gravity model would allocate the increase in production to western and central Canada, 80 per cent and 20 per cent respectively. This initial allocation puts a spatial dimension to the production increase. Soil erosion estimates can then be made using a revised version of the Universal Soil Loss Equation given the soil types, management practices and crop mix of the regions. These estimates could be adjusted to account for changes in management and production practices.

One of the advantages of using this method to estimate the environmental impact of a change in policy is its simplicity. Coefficients could be estimated and used for a variety of analyses. In those cases where the change in policy would have a major impact on a particular region, then the environmental coefficients would have to be modified to take this into consideration.

Incorporating the environmental impact into the PEM framework using this simple approach should be used with caution. The environmental coefficients will have to be modified with changes in either the bio-physical relationships or producer's behaviour. Information from the generic policy analysis will suffer from an aggregation problem if average environmental coefficients are used in the analysis. This approach can only be used when the bio-physical relationships are well known and baseline information is available. It could not be used where this type of information is lacking, for example biodiversity, or habitat availability.

4.   Specific policy response analysis

Analysing specific agricultural policies requires more detailed information. It also allows one to incorporate additional detail into the environmental impact analysis. Specific policy responses enable one to identify those AEIs that are most appropriate for the policy being considered. Information generated through this process can assist in clarifying policy objectives and target resources to areas of highest priority. It also provides a means of monitoring progress over time and allows for adjustments to be made to policies that will effect the future state of the environment. In addition, this information can also assist in programme design and implementation, ensuring that the environmental objectives are being satisfied.

The OECD (1997b) conducted a study to estimate the environmental effects of agricultural land diversion programmes in Canada, the European Union, Japan, Switzerland, and the United States. Elements of the land diversion programmes that had the greatest influence on the environment were: which land was set aside, the length of the set aside (short- or long-term) and what could be done with the land while it was in the set aside. Short-term land diversion programmes were usually introduced as a means of supply control and had a much smaller impact than longer-term programmes. Long-term programmes tended to be designed for environmental goals. All of the programmes analysed had voluntary producer participation and as a result producers were compensated for their participation. This had budgetary implications for the countries involved. Finally, the study concluded that AEIs should be developed in order that analysts can evaluate the environmental impact of these programmes. Without a set of AEIs, the study was limited in its ability to quantify the environmental impact, either good or bad, of these programmes.

5. Agri-environmental modelling

AEIs can be incorporated in a variety of modelling situations to increase the quantity and quality of information that can be used to evaluate detailed policy situations. The approaches discussed below give examples of how AEIs can be integrated into new or existing policy models. The types of policy analysis described falls in a continuum ranging from specific ecological models, sub-national and national models, macroeconomic models to multinational impact models.

5.1 Ecological level models

One approach for generating environmental information that can be used in the decision-making process, is to start by taking into consideration the ecological dimension of the problem to be analysed. Catchment scale models can evaluate water use and be aggregated to jurisdictional levels to evaluate sub-national policies (Russell, 1998). An example of this approach is the modelling of water resources in Australia's southern Murray-Darling Basin. The model uses simulation and optimisation techniques to evaluate policy changes related to water use. The modelling system is based on the physical flow of water through the catchment and the management of the water with dams for competing end users. One of the water uses in this area is for irrigation. Producer decision modules investigate the factors that affect producers willingness to buy or sell water. These factors include crop mix, commodity prices and seasonal variability of water (ABARE, 1998).

Another example of this approach to policy analysis is the generation of watershed scale models. In these cases the bio-physical relationships within the watershed are modelled in great detail. Dissart (1998) used this approach to undertake an economic evaluation of erosion control measures in the Saint-Esprit watershed in Quebec, Canada. A Geographical Information System (GIS) model was used to manage the bio-physical data related to the watershed. Producer behaviour was modelled with a Mixed Interger Linear Programming (MILP) model at both the individual farm and watershed levels. Erosion estimates were made using the Revised Universal Soil Loss Equation For Application in Canada (RUSLEFAC). Producer decisions on crop mix and tillage practice had a direct impact on erosion estimates. The model was used to simulate the impact of policies that would limit soil erosion in the watershed (Dissart, 1998). These policies were simulated by constraining the amount of soil erosion that was permitted within the watershed and allowing producers to select crops and tillage practices that would maximise producer returns within these constraints.

The advantage to this modelling approach is the quality of the bio-physical data that the model is based upon. The main disadvantage stems from the substantial data collection exercise involved and the difficulties associated with aggregating watershed or catchment estimates to other jurisdictional levels. Evaluating national policies over a large physical area would be problematic with this approach.

5.2 Sub-national and national models

A variety of sub-national and national models exist to analyse agricultural policy. These include both mathematical programming and econometric based models. Several government agencies have integrated AEIs into their modelling frameworks in order to capture the environmental impact of public policy. Sub-national models provide an excellent means of adding the scale dimension to environmental analysis. One of the important factors in understanding the impact of agricultural policy on the environment is to identify and quantify the bio-physical relationships. Often these relationships have a regional or sub-regional dimension to them and, as a result, sub-national models can provide additional information to the analysis of agricultural policy. This is particularly important when trying to design agricultural policies that are targeted to deal with particular environmental problems in particular areas.

Canadian Regional Agricultural Model (CRAM)

A number of OECD countries have existing sub-national agricultural policy models that can be modified to incorporate agri-environmental indicators. This type of modification integrates an economic policy model with a bio-physical model. One attempt at doing this is the CRAM-BIO-PHYSICAL modelling system developed by Agriculture and Agri-Food Canada (McRae *et al.,* 1995). The Canadian Regional Agricultural Model (CRAM) originated as a mathematical programming model of the prairie grain-livestock economies in the late 1970s. The model can be used to estimate the production decisions of crop and livestock producers, as market prices and policies change. The bio-physical model that was integrated with CRAM for Western Canada was the Erosion Productivity Impact Calculator (EPIC) which can be used to estimate wind and water erosion. For Central and Eastern Canada, the revised universal soil loss equation (RUSLEFAC) was used. The integration of these bio-physical and economic models required a number of assumptions with respect to scale.

The development of this integrated model was based on using the CRAM model as the agricultural producer decision component and having these decisions feed into EPIC or RUSLEFAC to estimate the environmental impact. CRAM was modified to increase its resource sensitivity to such things as tillage practices and producer risk functions. The output from the CRAM-BIO-PHYSICAL modelling system takes into account the risk associated with the producer, i.e. risk averse or risk neutral, and the risk associated with the environmental indicator, i.e. soil erosion risk class low, tolerable and high (Agriculture and Agri-Food Canada, 1995). Examples of the types of output that can be generated from this modelling approach are given in Tables 6 and 7. This model is currently being used to assess the environmental and economic impacts of Federal and Provincial crop insurance programmes.

Table 6. Change in land use as a result of a policy change (percentage terms)

Region (Sub-national Region)	Scenario 1 Highly Risk Adverse	Scenario 2 Moderately Risk Averse	Scenario 3 Slightly Risk Averse	Scenario 4 Risk Neutral	Lower Crop Returns
A 1	%	%	%	%	%
A 2	%	%	%	%	%

Source: Agriculture and Agri-Food Canada, 1995.

Table 7. Water erosion results from a change in policy (tonnes/hectare/year)

Region (Sub-national Region)	Soil Erosion Risk Class	Physical Change	% Change
A 1	Low	X	%
A 2	Tolerable	X	%
	Moderate	X	%

Source: Agriculture and Agri-Food Canada, 1995.

The objective function in the CRAM-BIO-PHYSICAL model is profit maximisation. As a result, producers choose that combination of crops, resources, technology and purchased inputs, based on relative prices, government programmes and risk, which is consistent with this behaviour. This behaviour permits a producer to substitute purchased inputs, crop mix and technology as relative prices change. These production management decisions will have resource impacts that can be estimated with the bio-physical models. EPIC and RUSLEFAC models are then used to estimate soil, water and wind erosion for the various sub-regions in the model.

One of the advantages of the CRAM-BIO-PHYSICAL model is that it is a fully integrated economic-bio-physical model. Changes in policy are directly integrated into the producer behaviour model, and this impact on production decisions produces environmental consequences. The model also provides a risk assessment of the policy change both in terms of economic behaviour and environmental resources. The output from the model can be presented in tabular or map form. The latter makes interpreting the results easier for analysts, stakeholders and decision makers. Finally, the output from the model can be provided at a number of scales, by province and sub-province. This can be an important attribute for decision makers when scale, both in terms of sub-national regions or environmental sensitive areas, is important.

Some of the disadvantages of the CRAM-BIO-PHYSICAL modelling system are the complexity of the model and the data requirements to maintain and up-grade the model features. In its current stage of development, it can only produce estimates for a limited number of AEIs, i.e. soil erosion, wind erosion and GHG emissions. The model is also limited in terms of the coverage of the sub-national regions in the country. Finally, the model takes into consideration only the direct impacts on the agriculture sector. It does not take into account other industrial sectors that directly provide inputs into the agriculture sector, or the environmental impact of the agri-food sectors.

Other modelling developments

In the future, Geographical Information Systems (GIS) will be fully integrated into sub-national models to highlight the spatial impact on the resource and environmental base to changes in agricultural policy. GIS can be used for data management and will provide decision makers with easy-to-interpret visual output. Currently, the CRAM-BIO-PHYSICAL model has some map generating capacity. Another example of a modelling system that uses GIS data management technology with economic and environmental modelling capabilities at the sub-national level is the Resource and Agricultural Policy Systems (RAPS). This modelling system was developed by the Center for Agricultural and Rural Development (CARD) at Iowa State University (CARD, 1998).

RAPS uses three components to estimate the environmental impact of policy change. *First*, site specific characteristics are used to provide information on resource characteristics, e.g. soil characteristics and climatic information. This information is linked to the *second* component, the Acreage Response Modelling Systems (ARMS) which is a producer decision-making model. ARMS predicts the choice of crop and tillage system, whether no-till, reduced-till or conventional till. These choices are based on probabilities which are obtained from site specific historical information. *Third*, this producer decision model is linked to a bio-physical model, the Site-Specific Pollution Production modelling system (SIPP), to estimate the environmental impact of producer decisions (Babcock *et al.,* 1998).

SIPP uses environmental production functions to estimate a number of environmental indicators. These fall into a number of broad categories: soil (e.g. water and wind erosion, soil organic carbon change), pesticide (atrazine run-off), and animal waste (nitrogen run-off, leaching and volatilisation). Erosion estimates are made with the Erosion Productivity Impact Calculator (EPIC) while the estimates of pesticide pollution are made with the Pesticide Root Zone Model (PRZM). Animal waste is estimated using animal population numbers, average weight classes and the average amount of waste generated per animal unit (Babcock *et al.*, 1998). An example of the type of information that is generated from this model is given in Table 8. The model estimated that the net change in soil erosion over the period 1992 to 1997 for Illinois, United States, was nearly 21 million tonnes. Of this total, over 5 million tonnes of erosion occurred because of crop changes while over 15 million tonnes were caused by changes in tillage practices.

One of the advantages of the RAPS modelling system is the integrated nature of the site-specific characteristics, the producer decision-making model and the bio-physical model. This model includes a large number of agri-environmental indicators that can be used for policy evaluation. Finally, the site-specific nature of the modelling system allows the analyst and decision maker to investigate the scale properties of the environmental impact. The use of GIS allows information to be generated on maps that provide easy access to the spatial dimensions of the problem.

Table 8. Soil erosion estimates in the United States Corn Belt

(1992 to 1997, million tonnes)

Region	1992	1997	Crop Change	Tillage Change	Net Change (%)
Illinois	166.61	187.28	5.34	15.32	20.6 (12.4)
Indiana	71.02	61.95	3.03	-12.11	-9.1 (-12.8)

Source: Babcock *et al.*, 1998.

One of the disadvantages of this modelling system is the cost of obtaining and managing its database. A large amount of data has to be generated before such a modelling system can estimate the various coefficients in the economic and bio-physical models. Another disadvantage is the number of assumptions that have to be made between the actual data and how it is used in the model. Many of the site characteristics are generalised over various scales of the analysis.

Regionalised Agriculture and Environmental System (RAUMIS) Model for Germany

Another example of a sub-national/national integrated model is the Regionalised Agriculture and Environmental Information System for Germany (RAUMIS) (see Meudt, 1999). RAUMIS is a supply model for agriculture that has detailed regional information for 431 regions and is composed of both a base model and a simulation model. The base model is used to calibrate the simulation model. The simulation model is a non-linear programming model that has an ecological component built into it. The objective function is to maximise net returns given the agricultural policy scenario, and production and environmental constraints (Meudt, 1999).

One of the unique aspects of this model is its environmental component. It includes a nutrient balance module, a biodiversity indicator, a landscape diversity indicator and the start of a greenhouse gas inventory. The nutrient balance can be estimated for nitrogen, phosphorus and potash based on the agricultural production activities undertaken. Nutrient input and removal are estimated for the cropping activities to provide an estimate of the net surplus, which is used as an indicator of the "potential pollution" from the activities undertaken (Meudt, 1999). This component takes into account the intensity of agricultural production and manure management strategies from the livestock components of the model.

The biodiversity indicator uses a set of 5 criteria to determine the impact that crop production and intensity have on biodiversity. The criteria were developed by 30 experts in the ecological sciences and include such things as closeness to the natural vegetation, impact on endangered species, and time to regenerate after the man-made disturbance (Meudt, 1999). Each expert provided a weight for each criterion in order that a composite index could be estimated.

The landscape diversity index provides an indication of the diversity of land use within a region. An index of land use is created using the number of hectares in production and the distribution of those activities in the region. The index increases as the number of crop activities increase, and as the share of land going into any activity becomes more homogeneous (Meudt, 1999).

The final environmental component is the development of a greenhouse gas emission module. It is being designed to estimate the greenhouse gas potential of the agriculture sector. Emission estimates from livestock activities are used to estimate the greenhouse impact given the level of agricultural activity (Meudt, 1999).

Meudt (1999) reports the results of a comparative static analysis using RAUMIS to simulate the impact of a liberalised Common Agricultural Policy (CAP) on the environment. In 2005, the model results indicated that cereal production would increase while oilseed production in Germany would decrease. While the average nitrogen surplus would be similar to the baseline, the distribution of the surplus differed. With liberalised trade the nitrogen surplus would decrease in eastern German. The biodiversity indicator improved with less intensive production. The land use diversity decreased with the increase in fallowland (Meudt, 1999).

United States Mathematical Programming System (USMP)

Another example of a national-interregional policy model that has been integrated with a bio-physical model is the United States Mathematical Programming system (USMP) (House *et al.*, 1998). USMP is a mathematical programming model that contains 45 regions and includes costs of production for both crop and livestock activities. USMP is integrated with the EPIC biophysical model to estimate the environmental impacts of policy changes. The model is also linked to a GIS system, Arc View, which can be used to map the results of the analysis. The economic model is calibrated to US Department of Agriculture's agricultural baseline, while the bio-physical calibration is a function of specific regional soil types, weather and cropping systems. The objective function in the model maximises profits subject to policy requirements and/or incentives (House *et al.*, 1998).

The model was used to estimate the economic and environmental impact of reducing national nitrogen use by 10 per cent. It was estimated that an ad valorem tax of 69 per cent would be required to reduce national nitrogen use by 10 per cent, and that this would increase the cost of production by 8 per cent, decrease acreage planted by 2.5 per cent and decrease farm income by 3 per cent (House *et al.*, 1998). The distribution of these impacts varies depending upon the region. Similarly, the environmental impact had an important distribution dimension. Nitrogen use declined by 3 per cent in the Mountain and Delta region and 18 per cent in the Southeast but nitrogen loss was only 8 per cent with declines of 1 per cent in the Delta States and 14 per cent in the Southern Plains (House *et al.*, 1998). The authors concluded that an ad valorem tax is not an efficient means of reducing nitrogen applications and leaching in those regions where it is a significant problem.

5.3   *Economic-ecological input-output models*

AEIs can also be integrated into existing input-output models. Input-output models can be used to estimate the macroeconomic impacts of changes in government policy. Traditionally, the output from these models will measure the direct, indirect and induced effects on industrial output, GDP and employment, resulting from a change in final demand. Augmenting this model with a set of environmental indicators will allow estimates of the environmental impact of a change in final demand.

Input-output models are general equilibrium models and can be developed at the national and sub-national levels. The models are based on information that is generated from the System of National Accounts. A brief description of the economic-ecological input-output model developed for Agriculture Canada (Thomassin and Carpentier, 1993) will be provided to give an indication of the information that can be generated from this type of model.

A rectangular accounting framework was used in the construction of this model. The number of commodities in the model is greater than the number of industries. With this framework each industry has the opportunity to produce more than one output. The accounting framework provides an estimate of the supply and disposition of the dollar flow of goods and services in the Canadian economy. It values the inputs going into the production process for each industrial sector, and values the outputs being produced by each sector. The model was based on Statistics Canada's National Input-Output Model. The agricultural sector in the Statistics Canada model was disaggregated in order to provide greater detail for agricultural policy analysis.

The model is shocked with a change in the final demand for commodities. The model will estimate how much industrial output will have to increase to satisfy both the direct increase in demand and the indirect increase required to produce all of the necessary inputs required for the production process. From the estimated changes in industrial output changes in GDP and employment can be estimated.

The accounting framework in the Agriculture Canada model was augmented to include a number of ecological commodities that are inputs or outputs of the production process. The ecological inputs and outputs were incorporated into the model on an industrial sector basis. Much of this information was already available. The ecological inputs going into the model fall into two broad categories, soil and water, while the ecological output fall into 5 categories: solids, waterborne, airborne, pesticides and fertilizers. A detailed list of the ecological commodities included in each category can be found in Table 9. Using this model, one can estimate the change in ecological commodities that result from a change in the final demand for commodities.

The following approach would be used to undertake the analysis. The change in final demand would have to be estimated initially, typically with the assistance of another economic model that estimates the change in agricultural output. This change in final demand is used to shock the model to estimate changes in industrial output. The estimated changes in industrial output are then used to estimate changes in GDP, employment and the ecological inputs and outputs.

Table 9. Ecological commodities included in the Economic-Ecological Input-Output Model

Ecological Inputs	Ecological Outputs
* Soil	* Solids
– Erosion	– Liquid Animal Waste
– Total	– Solid Animal Waste
– Delivered Erosion	– Solid Wastes
* Water	* Waterborne
– Industrial	– 5-Day Biological Oxygen Demand
– Total Intake	– Oil
– Total Recycled	– Suspended Solids
– Gross water Use	– Total Dissolved Solids
– Total Discharge	– Chemical Oxygen Demand
– Total Treated Discharge	– Nitrogen
– Agricultural	* Airborne
– Total Intake	– Nitrogen Oxide
– Livestock Watering and Irrigation	– Volatile Organic Compounds
	– Hydrocarbons
	– Sulphur Dioxides
	– Carbon Monoxides
	– Particulate Matter
– Municipal	* Pesticides
– Total Intake	– Phenoxy
– Total Discharge	– Dicamba/Bromoxynil
	– Triazine
	– Other Herbicides
	– Fungicide
	– Captan
	– Insecticides
	– Other
	* Fertilizers
	– Total Quantity
	– Phosphorous
	– Potash
	– Nitrogen

Source: Thomassin and Carpentier, 1993.

The development of this type of modelling framework has a number of advantages. Among these is the provision of estimates of both the economic impact of policy change, in terms of changes in industrial output, GDP and employment; and the environmental impact, in terms of changes in ecological commodities (expressed in physical terms). This provides analysts, decision makers and stakeholders with estimates of the trade-offs between industrial development and environmental consequences. An additional advantage to this approach is that it takes into account not only the direct effect of changes in the agricultural sector, but also the indirect economic and environmental impacts of the other industrial sectors that supply inputs into the agricultural sector. This can be of particular importance if countries adopt policies that shift the output of the agricultural sector from primary agricultural products to more processed agri-food products. This approach to evaluating the environmental impact of alternative policies provides a more complete picture of the environmental effects of changes in agricultural production. Finally, this approach allows decision makers to compare the economic and environmental impacts of changes in public policy across a number of industrial sectors.

This approach to modelling the environmental impacts of policy change has some disadvantages. *First*, the model is static and does not provide a timeframe over which the impacts will occur. *Second*, the change in final demand that is used to shock the model has to be estimated by the analyst and may require the use of other economic models. This will require the analyst to have some understanding of both this modelling framework and the other that must be used. *Finally*, the development of such a modelling system will require a large amount of resources if existing input-output accounting frameworks and environmental data do not exist. However, if these "disadvantages" can be overcome, decision makers will have information that takes into consideration social, economic and environmental dimensions of the sustainable development objective.

5.4   *Multinational impact models*

A number of environmental problems between countries occur because of the transboundary nature of the environmental impact. In the future more analysis will have to be undertaken in order to be able to take these problems into consideration. Examples of these transboundary pollution problems include climate change, acid rain, ozone pollution and cross border river pollution. AEIs can be incorporated into multinational impact models to provide information on the state of the environmental impact and on the effect that policy changes will have on the future state of the environment.

An approach that can be used to analyse transboundary problems is the integration of appropriate national or sub-national models from the countries concerned into a single model. The policy impact on each country's producers' production decisions would have to be modelled and linked into a bio-physical model that takes into account the transboundary pollution problem. Links would also have to be developed between the policies in each country and the neighbouring producer decision model.

Another approach is to design a multinational regional model. Examples of these types of models include the Regional Acidification Information and Simulation Model (RAINS) developed by the International Institute for Applied Systems Analysis (IIASA) (see Amann, 1993) and the Future Agricultural Resources Model (FARM) developed by the USDA (see Darwin *et al.,* 1996). The RAINS model estimates the economic and environmental impact of transboundary air pollution, in particular acid rain for various regions: Europe and Southeast Asia. Model simulations provide estimates of environmental and economic impacts for the various countries in the region. FARM has been used to estimate the impact of changes in climate, population and international trade on global land use and cover. This model estimates both global and regional impacts for the scenarios developed (Darwin *et al.,* 1996). These models provide decision makers with information on the environmental impacts in their own country as well as the impact on other countries.

5.5   *Multiple criteria decision-making models*

The modelling tools described above provide information to decision makers in terms of the economic and environmental impacts associated with particular policy changes. The models do not, however, provide a mechanism for assisting decision makers with analysing the trade-offs between social and economic benefits and environmental damage. Integrating such a mechanism into the informational framework would facilitate the decision-making process.

Evaluating policy in terms of sustainable development objectives requires decision makers to take into account several objectives of a social, economic and environmental nature, making evaluation of the trade-offs between the various objectives a very complex problem. A potential way to assist decision makers is to develop models that incorporate multiple criteria decision-making techniques (Romero and Rehman, 1987), an example of such a model is *compromise programming*.

Compromise programming recognises that in many situations the various objectives of the decision maker are in conflict and not all of the objectives can be satisfied. Compromise programming provides a framework for decision makers to select a second best solution when no Pareto optimal solution is available. The model is designed as an iterative process where the decision maker responds to the model solutions that the analyst presents. Through this iterative process trade-off curves are developed between objectives and a best compromise solution can be identified.

An example of a compromise programming application to agriculture is given by Romero *et al.* (1987). In this situation the multiple objectives included employment, seasonal labour and business profitability. An iterative filtering process was used and a compromise between the objectives was arrived at.

This approach has particular applicability when policies are being evaluated in terms of their sustainability, and thus have social, economic and environmental objectives. Multiple criteria models provide decision makers with a framework where they actively participate in identifying the best compromise across the various objectives. This integrates the policy decision maker into the on-going modelling process and develops a feedback mechanism between the producer decision-making model, policy decision maker and the bio-physical model.

6.   Non-modelling approaches

All of the above approaches rely on the construction of quantitative models that require a substantial investment in resources dedicated to the collection of data, management of databases, and model design. Thus, while these approaches provide decision makers with quantifiable estimates of the economic and environmental impacts of policy, there are substantial costs associated with data acquisition. There are other non-modelling approaches that can provide an indication of the environmental impact.

6.1.   *Qualitative assessment of policy impacts*

An information framework can be designed that identifies the potential impact of policy changes in a qualitative manner, and provides a systematic review of the potential environmental impacts. Once potential impacts have been identified, the analyst could provide a qualitative assessment of the likely risk and severity of the environmental damage, based on their knowledge of the agriculture sector being analysed.

Campbell (1998) gives an example of this type of information framework, designed around three Worksheets. The *first Worksheet* identifies the potential impacts that policies may have on producer production decisions (Table 10). The first column in the table identifies the type of programme that is being analysed while subsequent columns identify potential areas where the policy can impact on producer decisions.

The *second Worksheet* provides an indication of the environmental risk associated with the production decision (Campbell, 1998). The first column takes the production decisions information from Worksheet 1 and identifies potential environmental risks associated with it (Table 11).

Table 10. Worksheet 1: Impacts of policies on producer decisions

	Examples of Producer Production Decisions that may be Affected by Policy				
Type of policy being analysed	Land allocation	Crop selection, tillage, rotation	Nutrient application	Soil conservation	Wildlife management
Income stabilisation	?	?	?	?	
Crop insurance	?	?	?	?	?

Note: ? = Indicates a potential impact on a producer's production decision.
Source: Campbell, 1998.

Table 11. Worksheet 2: Impacts of producer decisions on environmental risk

	Examples of Environmental Risk Associated with Production Decisions				
Producer decisions from Worksheet 1	Soil erosion	Soil compaction	Soil salinisation	Sedimentation of water	Nutrient contamination of water
Land allocation	?	?	?	?	
Crop selection, tillage, rotations	?	?	?	?	?
Nutrient application					?

Note: ? = Indicates a potential environmental risk.
Source: Campbell, 1998.

The *third Worksheet* identifies the significance of the environmental risk (Campbell, 1998). The environmental risks identified in Worksheet 2 are placed in the first column of Worksheet 3 and an assessment of their significance is provided (Table 12).

Table 12. Worksheet 3: Significance of environmental risk

Insert Environmental Risks from Worksheet 2	Comments Indicate Ways of Determining Significance of Environmental Risk			
	Is the effect of large magnitude?	Does the effect cross critical thresholds?	Is the effect long-term or irreversible?	Is the effect of high public concern?
Soil erosion	Area affected, Quantity	Tolerable erosion rate?	Severe erosion is difficult to reverse?	
Soil compaction	Area affected	?	Deep compaction is hard to reverse	

Source: Campbell, 1998.

One of the advantages of this informational framework is that it provides a systematic way of analysing the environmental impact of policies, and has minimal data requirements. Much of the information that is needed to make the assessment can be obtained from published sources or personal experience. The time required to conduct the analysis will be shorter than most quantitative modelling approaches.

A disadvantage of this approach is the qualitative nature of the analysis. It is possible that if two individuals undertook an analysis of the same policy, they could arrive at different conclusions. Another disadvantage is that it does not give quantifiable estimates, or provide any detail on the time and space dimensions of the environmental impact. Problems can also arise when the environmental impacts are both positive and negative, when the critical question is to determine the net effect. Finally, the analysis does not address the social or economic dimensions of the policy analysis.

6.2   *Incorporating AEIs for other policy evaluations*

A number of producer organisations have indicated an interest in adopting voluntary measures for environmental protection. An example of this is the International Standard Organisation (ISO) 14000 standard for environmental management. The identification of AEIs as part of the monitoring and auditing process of ISO 14000 would provide producers and government agencies with additional information on the state and response of environmental management in the sector (Lussier *et al.*, 1997). This provides a natural link between the information used on-farm by producers and that used by public policy decision makers at the national level.

An advantage to this approach for assessing producer behaviour to policy is that the organisations that would certify producers could also collect the data on the AEIs. It would also provide information on the priority given to environmental concerns by producers, and the actions that they are taking to correct them. A disadvantage of this approach is the reliance on other organisations to gather and monitor the data collection process and the consequent danger that the information collected may not be relevant to the policies that are being proposed or are under review.

145

7. Integration of models to predict future impacts

Decision makers require information not only on the historical impact, but also on the future impact of current policies, or changes in policies, as well as on producer decisions and the environment. Many countries currently have the capacity for estimating medium-term future commodity outcomes. Examples of this type of analysis include OECD's Agricultural Outlook, (OECD, 1999); the Food and Agricultural Policy Research Institute (FAPRI) Outlook, Missouri, United States; the IMF World Economic Outlook; and Agriculture and the Agri-Food Canada's Medium Term Policy Baseline. All of these outlook analyses use a variety of models to predict future prices, agricultural commodity output, agricultural trade and farm income. The information generated from these outlook predictions can be used in conjunction with other models to predict the environmental impact of future policy changes.

The one common element of the models described in previous sections, is that they each have an economic decision-making model integrated with a bio-physical model. The models estimate the environmental impacts that result from producer decisions concerning crop mix, tillage method, etc. The economic decision-making component integrates the information from predictive models, that is outlook analysis, with the agri-environmental models in order to provide decision makers with an estimate of the future environmental impact of policy decisions.

This integration of outlook analysis with agri-environmental models is often not as simple as it might first appear. One of the potential problems is the compatibility of outlook estimates with agri-environmental models. Often the estimates that are required for these models are not the same as those generated from the outlook analysis, for example, outlook estimates are often at a higher level of aggregation than is relevant for site-specific environmental outcomes. Other potential problems include the applicability of agri-environmental models over time. Many of these models have been designed for static analysis or have underlining assumptions that may limit their use over extended time horizons.

In order to increase the complementarity of these modelling efforts a number of things can be done. *First*, analysts should be cognisant of the information that is generated from these models and take this into consideration in model design. *Second*, efforts should be made to develop interface matrices that will take estimates from these predictive models and convert them into a form that is usable for the agri-environmental model. The development of these interface matrices will improve the efficiency of the information transfer from the various models, and will minimise the potential for data transfer errors between models.

Increasing the complementarity between modelling systems will provide decision makers with additional information that will facilitate the decision-making process. For example, Agriculture and Agri-Food Canada uses the Food and Agriculture Regional Model (FARM) to forecast changes in prices and output of the Canadian agriculture and food sectors. These estimates could be used in the CRAM-BIO-PHYSICAL or Economic-Ecological Input-Output models to estimate the environmental impact of these changes at both the regional and national level, for both the agriculture sector and the other sectors in the Canadian economy. This information would provide both sector-specific and macroeconomic environmental impacts over time as well as a spatial dimension to the analysis.

8. Discussion

AEIs can play an important role in assessing the sustainability of public policy. Several countries have passed legislation that requires government agencies to undertake periodic environmental assessments of their policy. Examples are the Soil and Water Resources Conservation Act in the United States (USDA, 1989) and the Farm Income Protection Act in Canada. Other countries, such as The Netherlands, have developed comprehensive National Environmental Policy Plans that use a variety of indicators to set public policy (Hoogervorst, 1995). Establishing a set of internationally recognised AEIs would facilitate the evaluation of national policies and would have the added benefit of being understood by the international community.

Establishing a set of AEIs would also allow the periodic environmental evaluation by international organisations such as the OECD. Such a set of AEIs could be used to evaluate international agreements, for example trade agreements. Concerns have been expressed that the environment can be used as a strategic variable during and after trade agreements have been signed. For countries with high environmental standards, the requirements associated with those standards could be perceived as a non-tariff barrier to restrict trade or provide subsidies. Similarly, countries could be under pressure to decrease their environmental standards to increase their competitiveness.

Most of the recent international reviews of specific country's policies have depended upon the availability of data in that country, thus limiting their potential usefulness for international comparisons. Establishing a set of recognised AEIs would overcome this problem.

The recognition that public policy must be assessed for its environmental impact supports the development of modelling efforts that will generate quantitative information. However, a number of concerns have to be addressed in order that "quality" information will be generated. The first is the data requirements for model development. The models to be developed should be based on theoretical concepts that are relevant to the situation being considered, that enable proxies to be identified that represent these theoretical concepts and for which data can be collected and measured (Bonnen, 1987). Deficiencies in any one of these steps will result in data obsolescence. Currently, much of the data used in the economic analysis is generated separately from the environmental data (House *et al.*, 1998). As a result the linkages between economic and bio-physical data are not as strong as might be desirable.

Economists have had many years of experience developing policy models that estimate traditional policy variables such as prices, production and trade. The integration of bio-physical models with these economic policy models increases the level of complexity in the modelling process. As a result the feedback mechanisms between the decision maker and the environment and vice versa are not clearly defined.

The modelling of the bio-physical relationships is very difficult. In many cases the bio-physical models that are being developed are simplistic and lack the ability to predict environmental impacts accurately. Additional work is needed on the bio-physical models to increase their precision.

The current integration of economic and bio-physical models has concentrated on modelling the negative impact of agricultural production, for example, fertilizer and pesticide pollution or soil erosion. These bio-physical processes, though complex, are easier to model than environmental impacts related to biodiversity, landscape and habitat. Modelling these types of indicators would require a substantial input of resources.

Indicators need to be developed that have a "quality" dimension associated with them. These types of indicators would be used to identify those situations where the agricultural sector is providing environmental benefits to society that could be considered an environmental output. Part of this research effort will be to define threshold values that can be used as a reference point for production.

It is important that the integrated models that are developed endogenise the environment in the modelling process. In some of the current integrated models, the environmental impact is a residual calculation after the economic decision has been made. It is important that the environmental consequences of decisions are endogenised into the model with feedback mechanisms that impact on the economic decisions. Finally, models should be designed so that environmental policy is incorporated into the objective function and not simply viewed as a constraint that must be satisfied.

The development of integrated economic decision models with bio-physical models should be viewed as a decision support system that will provide useful and sound information for the decision-making process. Stakeholders and public decision makers can use this information to evaluate alternatives and to identify the trade-offs between the environment, social goals and economic growth. This should result in better resource and environment management and policy in the future.

Finally, the environmental impacts of some agricultural activities are so complex that information generated from the modelling process should be supplemented with an environment monitoring survey of sample areas. This information would be added to the decision-making process. Some of the issues that relate to environmental monitoring of sample areas include the timing and scale of the survey.

9.   Conclusions

AEIs are being developed in order to provide information to decision makers on the environmental impact of government policy. This is being done to assess the sustainability of current policies. As a result, analysts are developing new modelling frameworks that integrate both economic and bio-physical models.

The current trend in economic-environmental modelling is to develop models where policy influences producer production decisions, in terms of crop mix and tillage practices, and then to estimate the environmental impact of these decisions with bio-physical models. These new modelling frameworks are interdisciplinary in nature and require that economists, biologists and physical scientists work together to develop appropriate feedback mechanisms between the models. As a result, existing policy models are being modified to take into consideration this environmental dimension.

The type of information that will be required to assess existing policy will vary depending upon the policy. This will require that a variety of modelling approaches will have to be undertaken. The models that are currently being developed fall within a continuum in terms of their scale and complexity. Some models that have been developed are macroeconomic in scale and will provide information on the trade-offs between development and the environment at the national or international level. Other models are sub-national in focus and attempt to identify these trade-offs within a country, at regional or local level.

The following suggestions for future work are made in order to assist the development of modelling systems that will generate the necessary information for environmental policy analysis.

1. A recognised set of AEIs should be identified and agreed to. These should be developed by multi-disciplinary teams that will be using the AEIs in model development. AEIs should be evaluated in terms of their ability to provide useful information to policy decision makers.

2. Model development should proceed in conjunction with data development. Data sources should be reviewed in order to identify appropriate proxies that will be used in the modelling process. Both economic and bio-physical data should be collected and analysed together in order that the links between the economic and environmental activities are identified.

3. Additional resources are needed in the development of better feedback mechanisms between the economic decision-making model and the bio-physical model and vice versa. This research will increase the integration between economic and bio-physical relationships.

4. Models should be designed to endogenise both the economic and environmental aspects of policy.

5. Modelling efforts will have to occur at various levels of aggregation: macroeconomic, national and ecological dimensions. Models should be designed to interface with one another in order to capture the full range of environmental impacts.

In the longer term, models will have to evolve to adjust to the informational requirements of decision makers. In this case, more emphases will be placed on configuring model output than on the integration between economic and ecological processes (Hoogervorst, 1995). Mechanisms will have to be developed that link the informational output from the integrated models to decision makers. In this way decision makers will be proactive in addressing environmental problems before they occur.

BIBLIOGRAPHY

ABARE (1998), *Modelling Competing Demands for Water Resources in Australia: Southern Murray-Darling Basin*, Australian Bureau of Agriculture and Resource Economics (ABARE), on Internet at http://www.abare.gov.au/service/Publications/Water/murray.html.

AGRICULTURE AND AGRI-FOOD CANADA (1995), *Agricultural Policies and Soil Degradation in Western Canada: an Agro-Ecological Economic Assessment,* Policy Branch, Agriculture and Agri-Food Canada, Ottawa.

AMANN, Markus (1993), "Transboundary Air Pollution", *Options*, Winter 1993, pp. 4-9.

BABCOCK, Bruce, A.P. MITCHELL, T. CAMPBELL, T. OTAKE, P. GASSMAN, M. SIEMERS, T.M. HURLEY and J. WU (1998), *RAPS: Agricultural and Environmental Outlook,* Center for Agriculture and Rural Development, Iowa State University, Ames, United States, on Internet at http://www.ag.iastate.edu/card/divisions/rep/raps97/home.html.

BONNEN, J.T. (1987), "The Status of Agricultural Data Systems as the Basis for Policy", Paper presented at the symposium *Relevance of Agricultural Economics: Obsolete Data Concepts Revisited*, American Agricultural Economic Association meetings at Michigan State University, United States, 2-5 August.

CAMPBELL, I. (1998), *Guide to the Environmental Evaluation of Agricultural Policies and Programs*, Environment Bureau, Adaption and Grain Policy Directorate, Policy Branch, Agriculture and Agri-Food Canada, Ottawa.

CENTRE FOR AGRICULTURAL AND RURAL DEVELOPMENT [CARD] (1998), "Card Maps Agricultural Production and the Environment", *CARD Report*, Vol. 11, No. 1, pp. 1-3, Iowa State University, Ames, United States.

DARWIN, Roy, Marinos TSIGAS, Jan LEWANDROWSKI and Anton RANESES (1996), "Land Use and Cover in Ecological Economics", *Ecological Economics*, Vol. 17, pp. 157-181.

DISSART, J.C. (1998), *The Economics of Erosion and Sustainable Practices: The Case of the Saint-Esprit Watershed*, M.Sc. Thesis, Department of Agricultural Economics, McGill University, Ste. Anne de Bellevue, Quebec, Canada.

GARDNER, Bruce L. (1987), *The Economics of Agricultural Policies*, Macmillan Publishing Company, New York, United States.

HOOGERVORST, Nico J.P. (1995), "Integration of Economic and Ecological Modeling of Agriculture in the Netherlands", in *Integrating Economic and Ecological Indicators*, J. Walter Milon and Jason F. Shogen (eds), Praeger Publishers, Westport, United States.

HOUSE, Robert, Howard McDOWELL, Mark PETERS and Ralph HEIMLICH (1998), "Agriculture Sector Resource and Environmental Policy Analysis: An Economic and Biophysical Approach", invited paper at the symposium *Environmental Statistics: Analyzing Data for Environmental Policy* sponsored by the Novartis Foundation, May, London.

LUSSIER, George, Laurie BAKER and Paul J. THOMASSIN (1997), *An Environmental Management System for Agricultural Production in Quebec*, Final Report, Canada-Quebec Entente, Department of Agricultural Economics, McGill University, Ste. Anne de Bellevue, Quebec, Canada.

MEUDT, Markus (1999), "Implementation of Environmental Indicators in Policy Information System in Germany", Chapter 15 in F. Brouwer, and B. Crabtree (eds), *Environment Indicators and Agricultural Policy*, CAB International, United Kingdom.

McRAE, T., N. HILLARY, R.J. MacGREGOR, and C.A.S. SMITH (1995), "Role and Nature of Environmental Indicators in Canadian Agricultural Policy Development", in S. Batie (ed.), *Developing Indicators for Environmental Sustainability: The Nuts and Bolts,* Special Report (SR) 89, Proceedings of the Resource Policy Consortium Symposium, Washington, D.C.

OECD (1997a), *Environmental Indicators for Agriculture*, Paris.

OECD (1997b), *The Environmental Effects of Agricultural Land Diversion Schemes*, Paris.

OECD (1999), *The Agricultural Outlook 1999-2004*, Paris.

ROMERO, Carlos, Franscisco AMADOR and Antonio BARCO (1987), "Multiple Objectives in Agricultural Planning: A Compromise Programming Application", *American Journal of Agricultural Economics*, Vol. 69, pp. 78-86.

ROMERO, Carlos and Tahir REHMAN (1987), "Natural Resource Management and the Use of Multiple Criteria Decision-Making Techniques: A Review", *European Review of Agricultural Economics*, Vol. 14, pp. 61-89.

RUSSELL, Les (1998), Personal Communication, Land and Water Resource Division, Department of Primary Industries and Energy, Canberra.

THOMASSIN, Paul J. and L. CARPENTIER (1993), *Development of an Economic-Ecological Input-Output Model for Agricultural Policy Analysis*, Final Report, Agriculture Canada, Ottawa.

US DEPARTMENT OF AGRICULTURE (1989), *The Second RCA Appraisal: Soil, Water, and Related Resources in Nonfederal Land in the United States, Analysis of Conditions and Trends*, Miscellaneous Publication No. 1 482, Washington, D.C.

DEVELOPING AND USING AGRI-ENVIRONMENTAL INDICATORS FOR POLICY PURPOSES: OECD COUNTRY EXPERIENCES

by
David Baldock,
Institute for European Environmental Policy,
London, United Kingdom[1]

1.   Background

Within the OECD Member countries there is a great range of agricultural systems and practices, some changing rather rapidly, others following a traditional pattern. The relationship between these practices and the many ecosystems in which they are applied is diverse and often complex. The intricacy of the linkages between agricultural activity, the environment and policy interventions creates a demand for clarity, simplification and synthetic indicators. It also introduces substantial technical and intellectual challenges and certain hazards, such as over simplification.

The use of indicators as an aid to policy making in the agri-environment context is a relatively recent phenomenon with a marked increase in activity during the 1980s. It is still a developing field; a set of standard international indicators has yet to emerge. However, it can be seen as a recent stage in a much longer-term development of indicators in policy work, that has mirrored a trend towards increasing sophistication and growing public scrutiny of policy decisions. The new classical economic indicators of growth, unemployment and inflation were perhaps the first to become widely used and debated, in the early part of this century. With the establishment of new areas of welfare and education policy in the 1930s-1950s, a second wave of indicators of poverty, literacy, public health and other social variables began to assume prominence. Since the 1970s, growing environmental concern has generated an interest in devising and tracking a new range of indicators relating to environmental quality and degradation. Agri-environmental indicators are an example of a new generation of more specialised environmental indicators concerned with a particular economic sector, and absorbing some of the objectives of the sector alongside purely environmental issues. Typically, these new indicators attempt to focus on the relationship between agriculture, the policies that seek to guide it, and the effects of both policy and practice upon the environment.

In most respects, the development of agri-environmental indicators is a more ambitious task than that of creating core economic indicators, since it is necessary to cover a wide range of both natural and anthropogenic activities, with great variations between sites, regions and countries. The debate over agri-environmental indicators is informed by a growing, but far from complete, understanding of the fundamental relationships between agriculture and the environment. For some issues, the research,

---

1.   I am grateful to my colleagues Janet Dwyer and Karen Mitchell for their contributions to this paper.

underlying analysis, and availability of data are still at a relatively early stage of development and there are major differences between countries in the attention given to particular topics. Such factors inevitably influence the extent to which meaningful indicators can be developed or applied for policy purposes.

Both governmental and independent organisations have been active in the process of developing agri-environmental indicators, particularly since the United Nations Conference on Environment and Development (UNCED) Rio de Janeiro, Brazil, in 1992. Aside from academic interest, some of the reasons for developing indicators appear to include:

- encouragement by international bodies, such as the UN Commission on Sustainable Development (UNCSD), which has led to many countries preparing sustainable development strategies and considering the role of indicators;

- the need to monitor progress in meeting international commitments, such as the Biodiversity Convention and Kyoto Protocol;

- the growth of legislation embodying specific environmental goals and targets, some expressed in quantitative terms;

- increased interest by many governments in evaluating the environmental impacts of policies, employing both ex-post and ex-ante techniques;

- a trend towards legislation in which there is a formal obligation on public authorities to evaluate the impact of their policies (cf. Canada and the EU);

- their potential as a tool for stimulating dialogue and developing both research questions and policy options in national or regional sustainable development debates;

- finding a means of facilitating and encouraging public understanding and participation in debates about policy and management options, internationally, nationally and at a local and regional level;

- a growing perception that such indicators are relevant to the international debate on agriculture policy, support levels, international trade and environmental quality, especially in the World Trade Organisation (WTO).

Indicators have a variety of potential uses for policy makers. Many expect that they will contribute to the analysis, monitoring and measurement of agricultural impacts and to the environmental evaluation of policy measures and options. Indicators are widely seen as offering one tool for consistent analysis over the longer term, often using time series data and for reducing the number of parameters in play in the agri-environment debate.

In many quarters, expectations about the usefulness of indicators appear to be rising, even though most appear to be only at an experimental stage of development. The perspective of different actors in the policy debate can vary considerably, reflecting the practical, institutional and political difficulties involved in agreeing, isolating, reporting upon and interpreting the "key" elements of agro-ecosystem behaviour and policy. For example, some of those involved in trade aspects of the international policy debate appear anxious to agree indicators relatively rapidly in order to bring an environmental dimension into the evaluation of agriculture policy in the near future. By contrast, scientific specialists often express caution about premature or inappropriate compression of complex relationships into apparently simple indicators.

For most policy makers involved in governmental institutions, indicators must be clear and meaningful in relation to the objectives of their own organisation. In addition, they should be accessible, reliable and able to be compiled without heavy expenditure on further research or data collection. Indicators should be understandable and credible, not only within the department in which they originate, but also in other arms of government, so that they can be utilised, if necessary, in interdepartmental exchanges and negotiations. For example, indicators may help to provide a dispassionate analysis of the outcome of a particular agricultural policy. This may need to be presented to budget or economic ministries that often control the flow of funds to agriculture departments.

Public authorities are currently engaged in analysis and development at a range of geographical scales. Some are seeking to apply indicators to an individual region, but most are concerned primarily with the national scale and are interested in variations arising within national boundaries. Many are at an early stage of strategic environmental indicator development — aiming for "state of the environment" reporting — within which agri-environment indicators are not clearly distinguished from other rural indicators.

However, an expanding group is concerned with international comparisons of both strategic and specific agri-environmental indicators, where the work by the OECD has a central role. International comparisons help to validate and increase the accountability of national work, giving a measure of national performance. They are also relevant where countries are required to submit reports or meet targets under international conventions. Increasingly, however, the environment is perceived as an aspect of international competitiveness and a potential element in trade negotiations. For example, indicators may be used in an attempt to measure where there is an environmental justification for policies presented as being environmentally beneficial, but considered excessively production related and trade distorting by other governments.

The OECD Member countries have identified 13 issues where agri-environment indicators are potentially useful and work on developing them is underway, with some more advanced than others. Lists of issues of this kind or of more precise indicators, have been published in several countries, although their objectives and status varies considerably. Many are the product of research programmes, exploratory studies, consultation exercises and similar activities. In general, this is an area of "work in progress", rather than *fait accompli*.

This paper is concerned with the development and use of agri-environment indicators for policy purposes, although it is not always easy to separate experience within the policy community from more abstract or academic work which is free from policy constraints. It is based primarily on information supplied to the OECD Secretariat or direct to the author from the Member countries, and other material used to construct a preliminary overview of experience.

2.   The policy framework for indicators

Most policy makers concerned with agri-environmental issues at a national level are confronted with fragmented information from a variety of sources on several different issues. These can be difficult to harness into a robust framework for analysing different sets of policy choices. There is a potentially large range of factors inside and outside agriculture which affect key characteristics of the natural environment. Over time, there has been a tendency for the number of environmental issues within the policy frame to increase but the information required to link a specific farm practice with a particular environmental outcome does not always keep pace and may not be available at all in a reliable form. Where it is available at a farm or local level, its applicability at a larger geographic scale may be limited or in doubt.

In this context, it can be difficult to make a thorough evaluation of the impact of certain agricultural policies. Similarly, analysts may not be confident in estimating how far agriculture is contributing to, or detracting from, national sustainable development goals. Indicators provide a potentially useful tool in crystallising key questions, condensing the information available into a manageable form, and providing a benchmark for measuring progress or a target to aim at. Indicators are not always developed within an explicit framework, such as the Driving Force-State-Response (DSR) model proposed by the OECD. However, such a framework can provide clarity and consistency, help to position indicators within the appropriate place in the policy making cycle, and allow comparisons between different sectors where indicators are being used.

Many governments are beginning to invest more in environmental information and are concerned to use indicators and other tools for relating this to policy decisions in a systematic way. However, the pace of progress has been variable.

In the 1960s and 1970s some of the early work on indicators addressed questions about the state of the environment, the pace of change and the sources of adverse pressure. Typically, the focus was on the global or national environment and not on individual sectors, such as agriculture. Within the agricultural policy community, environment was not usually seen as an important concern. One exception was work on the evaluation of agricultural land and of soil erosion, which has a long tradition in many OECD countries and in the FAO. The FAO and the International Institute for Land Reclamation and Improvement in The Netherlands set up a joint project to develop an international framework for land evaluation in 1976. Eleven years later the *Global Assessment of Soil Degradation* (GLASOD) was launched by United Nations Environment Programme (UNEP), FAO and the International Society of Soil Science (Sombroek, 1997).

There is a continuing stream of work on environmental indicators of relevance to the whole territory and economy, usually led by environmental agencies, and this is an important element in the spectrum of agri-environment indicators. Two other major streams are of more recent origin. One, concerned with progress towards sustainable development, has emerged over the last decade, especially since UNCED in 1992. Agriculture is an important sector in Agenda 21 itself and in the work taken forward in many OECD Member countries. The second stream, which has also come to prominence in the 1990s, arises from the more specialised needs of those concerned primarily with agri-environmental issues. Agricultural policy makers, for example, have faced greater pressure to assess the environmental impact of current and proposed policies or to devise plans for meeting precise environmental targets for their sector. New policies, such as incentive schemes for environmentally appropriate farm practice have created fresh evaluation and targeting issues. Wider aspirations to link agricultural and environmental objectives have had to be focused, and priorities set, while established support policies have come under closer environmental scrutiny. As these specifically agricultural issues have begun to drive work on indicators, the range of subject matter has widened, for example to include farm management and financial issues.

In the light of this brief history, some of the main contemporary applications for indicators are outlined briefly below. They include:

- State of the Environment reports — for tracking progress and raising public awareness;

- sustainable development reports — for establishing key issues, tracking progress and raising awareness;

- a response to international agreements, conventions and initiatives;

- progress in meeting environmental policy and sustainable agricultural policy objectives;

- evaluating local and regional impacts of particular agricultural and agri-environmental policies;

- targeting and evaluating bids by participants in agri-environment schemes;

- longer-term predictive analysis of agri-environmental issues, sometimes using models.

2.1   State of the environment reporting

Many OECD countries follow the classical DSR model (or its variant the Pressure-State-Response [PSR] model) in compiling national reports on the state of the environment. Typically, these cover all aspects of the economy extending far beyond agriculture. Italy, for example, publishes such a report every two years and is creating a national environmental information and monitoring system. Many other countries either publish reports or submit information of a similar kind to international bodies such as the UNCSD. The usual pattern is to consider the overall state of the environment, including available data, indicating key changes over time and to include agriculture amongst the activities creating pressures. In some cases, agricultural policies, such as agri-environment incentive schemes, are listed amongst the responses.

These strategic reports are concerned with the overall state of the environment, not specifically the agriculture sector. Consequently, environmental indicators and targets tend to be cross sectoral and are not designed to inform a more detailed analysis of the relationship between changes in agricultural practice and environmental outcomes, or to illuminate the results of specific policies. For example, in New Zealand the Ministry for the Environment is developing a set of integrated Environmental Performance Indicators (EPI) in which it is acknowledged that any single indicator may be relevant to a variety of sectoral policy goals (Ministry for Environment, New Zealand, 1997).

A second important strand of strategic thinking and policy initiatives is the development of national plans or programmes for sustainable development. Generally, these have a wider scope than state of the environment reports and incorporate both social and economic elements to varying degrees. The Rio de Janeiro conference in 1992, and resulting agreement on Agenda 21, provided considerable impetus for such reports. Like State of the Environment reports, most are cross sectoral rather than being concerned specifically with agriculture. Broad indicators included in Agenda 21 are shown in the Appendix. Box 7 shows some of the relevant elements of the resulting process for the follow-up to the UNCSD special session in 1997.

Box 7. AGENDA 21: National reports on implementation for the UNCSD special session in 1997

For this special session, governments were asked to submit country profiles which contain a summary of national action on each of the Agenda 21 issues, along with supporting statistical data and indicators as appropriate. The issues relevant to agriculture and the environment are:

— managing fragile ecosystems: combating desertification and drought (Chapter 12);
— managing fragile ecosystems: sustainable mountain development (Chapter 13);
— promoting sustainable agriculture and rural development (Chapter 14);
— conservation of biological diversity (Chapter 15);
— protection of the quality and supply of freshwater resources: application of integrated approaches to the development, management and use of water resources (Chapter 18);
— participation by major groups at the national and local levels: strengthening the role of farmers (Chapter 32).

Statistical data/indicators listed in the UNCSD Country Profiles for the above include:

— changes in the area of land affected by desertification (km^2);
— changes in the area of agricultural land and the consumption of fertilizers in kg/km^2 of agricultural land;
— protected area as a percentage of the total land area and number of threatened species;
— freshwater availability and annual withdrawal of freshwater as a percentage of total available water.

There do not appear to be any indicators for sustainable mountain development or strengthening the role of farmers.

Some countries have set out specific quantitative environmental goals for agriculture policy in their submissions to UNCSD, such as targets for the area of land farmed organically. Others have referred to qualitative goals such as the prevention of rural depopulation or minimising the loss of rural land to development (in some areas).

Source: UNCSD, 1996 and national submissions by governments.

2.3   Responses to international agreements, conventions and initiatives

The OECD itself has been a significant initiator of research, development and co-operation at an international level, beginning with a programme of work on environmental indicators in 1990 in response to a request from the 1989 G7 summit. Both this programme and the activities of the OECD Joint Working Party of the Committee for Agriculture and the Environment Policy Committee have provided a stimulus for national endeavours and a framework which has proved influential in the agri-environment field.

Most international conventions on the environment are cross-sectoral, covering topics such as climate change, protection of the ozone layer, the control of trade in wastes, etc. As such they do not require the development of specific agri-environment indicators. However, they do generate obligations to gather data, submit regular reports and meet formal commitments, focusing and structuring several key environmental policy issues. The UN Framework Convention on Climate Change (UNFCCC), which came into force in 1994, and the Biodiversity Convention are notable examples. The UNFCCC not only creates obligations to control emissions of greenhouse gases, but has led to the assembly of inventories of data concerning these gases, including a substantial programme of work on the agriculture sector. The analysis of emissions, sinks and the farming activities associated with them leads naturally to the selection of indicators.

2.4   Measuring progress towards environmental goals and sustainable agriculture

National environmental policies and legislation are a growing force in setting frameworks for the development of indicators, although these are of variable relevance to the agriculture sector. Some legislation, such as the European Communities Nitrates Directive, is concerned specifically with agriculture and, in setting targets, can lead directly to the establishment of indicators, which may be simply the thresholds laid down in legislation. Targets affecting the agriculture sector, such as reductions in pesticide use, or manure surpluses, may not be legally binding but can play a similar role in structuring work on indicators.

Not many countries prepare environmental performance indicators annually but one exception is The Netherlands, where a set of indicators covering most aspects of the environment has been published annually since 1991, following the Pressure-State-Response model. The role of agriculture and other sectors as a contributor to various environmental pressures, such as the use of pesticides, is identified specifically.

It is notable that the concept of sustainable agriculture is becoming more influential in many countries and is driving some of the more recent work on agri-environmental indicators. For example, in the United Kingdom, the Ministry of Agriculture has recently issued a consultation document on the development of "a set of indicators for sustainable agriculture in the United Kingdom" (MAFF, 1998). In the introduction to this document, indicators are described as "a form of summary statistics which can help to highlight the main sustainable development issues for policy makers, business and the public". A similar approach is evident in the United States, France and Norway, where there are reports with a particular focus upon the effects of farming practice and farming change upon environmental variables.

These approaches have in common an attempt to measure and analyse environmental change at a national level. Often they seek to identify trends over time and they may be utilised as a tool for judging national performance, both by policy makers and by other interests. They are also often intended as tools for raising public awareness of environmental issues. However because of their relatively crude national scale, they have more limited use in promoting appropriate policy design.

2.5 *Measuring progress: local and regional indicators*

The national level is not always the most appropriate one for developing agri-environment indicators, especially as the problems of aggregation from farm, to regional, and then to national level can be very considerable.

The Economic Research Service of the United States Department of Agriculture is one of the agencies which has worked at both the national and regional level, for example targeting particular problem areas. For this reason, they are developing and using indicators which are geographically disaggregated and are sensitive to differences in the agricultural resources being used in regions within the country (Heimlich, 1995). Similarly, in many European countries it is increasingly common for national governments to present indicator data in mapped form, so that variations at regional or more local level are immediately apparent (e.g. IFEN, 1997a). This makes it easy to separate broad national issues such as agriculture's contribution to greenhouse gases from more localised ones such as areas of irrigated land, and to tailor policies appropriately.

2.6 *Indicators in policy evaluation and targeting*

A potentially narrower, but nevertheless important, role for indicators is in the evaluation of specific policies. In this framework, indicators need to be highly relevant to the objectives of the policy being evaluated, the way in which it is delivered, and its anticipated impacts, which may be indirect as well as direct. They may include a mix of environmental, agricultural, institutional and economic or accounting measures, so as to measure the efficiency as well as the effectiveness of the policy being evaluated. In this context, the policy itself must be seen as a key driving force, although in other contexts it may be regarded as a response. The indicators used in such evaluations may be classified as state or response indicators, although those which relate to policy efficiency (e.g. cost of implementation) are perhaps less easy to divide in this way.

For another group of policy makers, the PSR/DSR models may be less relevant as a framework. Their main concern is to find an effective way of targeting agri-environmental policy so as to produce the greatest return for the budget available. Indicators can be used for prioritising applications to agri-environmental schemes, typically by using a scoring system. For example, in the European Community and several other European countries, there has been a rapid growth in agri-environmental schemes offering farmers incentives for entering management agreements, typically over a five to ten year period. In some cases, the authorities administering such schemes have made use of indicators to allow them to judge the merits of different applications and to select the most beneficial schemes for participation. Similarly, in the United States, there has been a need to develop criteria for choosing the best applicants to participate in certain government programmes. In particular, the Conservation Reserve Program (CRP), which retires land from agricultural use, is now targeted on land which is considered to offer the greatest environmental benefits. This requires a system of classifying and ranking land put forward by farmers for enrolment during the current "sign-up" period (see Section 3.4).

2.7   Predictive modelling

In theory, once agri-environmental indicators have been in use for some time and their behaviour is established, they can be used in modelling work to predict the environmental impact of various policy options.   Practical examples of this kind of application appear   limited, probably because the development of agri-environmental indicators is mainly in its infancy. Work is reportedly underway in several Member countries including the United States, Austria, The Netherlands and Germany. As a crude measure of environmental impact, land use statistics have been available for many years and they have been used in predictive modelling work. In the United Kingdom, for example, policy makers have supported a long-term project to develop the environmental capacity of a national Land Use Allocation Model (LUAM) which can be used to predict how the balance of national land use may alter in response to policy changes. However, because the model has effectively treated the whole country as a single farm, it can only give crude indications of the scale of environmental impact nationally — for example, in predicting a national increase in the area of certain crops, whose precise environmental impact would depend greatly upon where, and how, such crops were grown.

In Canada, predictive modelling is one of the aims for the future of their agri-environmental indicator (AEI) project (see Section 3.2 for more details). The indicators being developed under this project will be used in a new generation of integrated models which are being built from existing economic and biophysical models already in use in Canada. Multidisciplinary teams of economists, scientists and policy makers will be involved. Recent work has focused upon connecting biophysical models for soil erosion to an economic land allocation model — CRAM — which is similar to the UK's LUAM (Office of Auditor General, Canada, 1997).

2.8   Summary

It is clear that the PSR and related DSR models have been influential amongst policy makers in OECD Member countries. This is perhaps the most widely used framework for indicators, particularly in cases where authorities are seeking to produce national reports on the environment, which identify changes over time and the principal pressures and forces accounting for them. The PSR approach is also applicable to policy makers producing national reports on sustainable development, or progress towards sustainable development. In both these contexts, indicators are being used at a fairly strategic level, e.g. to help raise public awareness about environmental issues and to keep track of general impacts. However, there is also a range of other more specific frameworks, including the policy evaluation framework in which indicators are beginning to be used, where the PSR and DSR models may have less to offer.

Where policy makers have adopted an explicit conceptual framework, it is likely to reflect their analytical and policy objectives. For example, in 1995, the Danish National Environmental Research Institute (NERI) proposed the use of a framework which distinguished driving forces, pressures, states, impacts and responses (Holten-Andersen *et al.*, 1995). This became known as the DPSIR framework and has since been more widely adopted by the European Environment Agency (EEA), for example in the EEA's State of the Environment reports (Stanners and Bordeau, 1995). The DPSIR provides a framework to investigate the opportunities for improved efficiency in the use of natural resources and evaluate the adequacy of policy responses to stresses on natural resources.

In short, indicators are being introduced into the policy making process in an *ad hoc* way in response to short-term policy pressures, as well as within a more formal framework of the kind developed by the OECD. Many of these pressures arise from new legislation and initiatives which have introduced requirements to undertake evaluations, meet specific targets or assist the actual operation of agri-environmental programmes.

Where the DSR approach is adopted, it is not always clear how to specify policies themselves, particularly agri-environmental policies, within the framework. On the one hand, these policies can be seen as driving forces, especially where the focus is on policy evaluation. On the other hand, policies are often classified as a form of response to an environmental issue and clearly have this role as well. This suggests that the classification has proven value as an aid to analysis, but it should be used flexibly and creatively rather than becoming a rigid structure potentially constraining the selection of appropriate indicators.

3.   Some examples of indicators used by policy makers

In the past, indicators were often developed by environmental or academic bodies, slightly detached from agricultural policy management processes. More recently, policy makers and other interest groups have become much more directly involved in identifying and using indicators, often as a direct result of the various initiatives following the 1992 Rio Conference, such as Biodiversity Strategies and Action Plans. The examples here are not exhaustive but have been chosen to illustrate the variety of purposes and processes that exist.

3.1   International work on indicators

Several international organisations are actively involved in the development of environmental or sustainable development indicators. Most of these do not apply specifically to the agricultural sector, but are concerned with a wider environmental or development spectrum. For example, the World Bank published an ambitious set of *World Development Indicators* in 1997 which included environment amongst a number of broad themes, such as the economy and global linkages (World Bank, 1997a). In partnership with UNEP, the United Nations Development Programme (UNDP) and FAO, the World Bank has been involved in another global initiative to produce "land quality indicators". These follow the classic PSR framework and are highly pertinent to agriculture, covering topics such as soil erosion, soil quality, soil salinity and alkalinity, pasture quality, agro-biodiversity, etc. This work is aimed particularly at developing countries and has revealed significant shortcomings in the available data (World Bank, 1997b; Dumanski *et al.*, 1998).

The UN Commission on Sustainable Development (UNCSD) has been taking forward a work programme on indicators of sustainable development since April 1995. This uses the Driving Force-State-Response approach and covers the whole spectrum of social, economic and environmental aspects of sustainable development, including some attention to institutional aspects as well. One of the first results was a major publication including approximately 130 indicators, each with a methodology data sheet providing guidance on its significance, the way it is calculated, the scientific background and potential data sources (UNCSD, 1996). Several of these are of direct relevance to agriculture (see Appendix). It is intended to reduce this present set to a smaller, more manageable group over time. It is hoped that individual countries will utilise this resource in building their own indicators. The initiative has undoubtedly stimulated work at a national level; 16 governments, including several OECD Member countries, have volunteered to test at least some of the indicators developed by the UNCSD.

One of the more ambitious initiatives to develop consistent environmental pressure indices for a group of OECD countries is a project for EU member States led by EUROSTAT (EUROSTAT, 1998). This arises from a political decision within the EU to develop indicators covering all aspects of the environment, together with "green accounting" tools in support of the Union's fifth Environmental Action Programme. The Programme is a strategic planning tool explicitly referring to several economic sectors, including agriculture. The project is intended to describe key pressures, causing harm to the environment in a systematic, comprehensive and comparable way. These are divided into ten environmental policy fields, including several of relevance to agriculture such as "Biodiversity Loss", "Air Pollution" and "Water Pollution and Water Resources". The aim is to publish a set of 60 indicators for at least 12 EU member States by autumn 1998. The 60 proposed have been selected from a larger group of 100 on the basis of advice from scientific experts. Several indicators are of direct relevance to agricultural activity, such as:

- emissions of nitrogen oxides, methane, carbon dioxide;

- loss, damage and fragmentation of protected areas;

- area used for intensive agriculture;

- wetland loss through drainage;

- change in traditional land use practice;

- eutrophication;

- nutrient balance of the soil;

- pesticides used per hectare of agricultural area in use.

3.2   *National examples of strategic assessment work*

So far there has been rather limited application of specifically agri-environmental indicators for formal policy purposes in OECD Member countries. In many countries, work to develop and track indicators remains largely within the research sector, without any direct link to the policy process. However, the results of such work may include reports on agri-environmental issues which can be used by policy makers.

In the *United States*, the Economic Research Service (ERS) of the US Department of Agriculture undertook a substantial research effort in the mid 1990s leading to publication of a handbook entitled *Agricultural Resources and Environmental Indicators* (USDA, 1997a). This covers a wide range of indicators and uses them to support a commentary on recent trends and environmental policy issues related to agriculture in the United States. ERS describes its mission as "to provide economic and other social science information and analysis for public and private decisions on agriculture, food, natural resources and rural America". In the handbook, six main categories of indicator covering:

- land — land use, tenure, land and soil quality, land values and taxes;

- water — water use and pricing, water quality;

- production inputs — nutrients, pesticides, energy and farm machinery;

- production management — crop residue management, cropping patterns, pest management, nutrient and water management;

- technology — development of agricultural technology;

- US government conservation and environment programmes — water quality, conservation reserve, conservation compliance and wetlands programmes.

The indicators used are a mix of basic physical variables such as pesticide use by weight of active ingredient on selected crops, more synthetic ones such as pesticide risk indicators, and purely economic ones such as pesticide prices. No PSR classification is used. Data is often disaggregated to show regional or inter-state differences, which are particularly pronounced within the United States, and there is a substantial commentary, including analysis such as "Lessons learned from Water Quality Programs".

In a few countries, effort is more specifically focused upon providing indicators for direct policy use. *Canada* is a leading example, where an Agri-Environmental Indicator (AEI) project has been underway since 1993 (Agriculture and Agri-Food Canada, 1998). It is due to report in 1999 and is linked to a modelling exercise with both economic and bio-physical elements. The major categories selected for AEIs are (see Appendix):

- soil degradation risk;

- risk of water contamination;

- agro-ecosystem greenhouse gas balance;

- agro-ecosystem biodiversity change;

- input use efficiency;

- farm resource management.

The Canadian approach is a highly technical exercise, whereby each of these categories is intended to be represented in a single indicator, but these will in turn be derived from a group of component indicators. For example, the indicator for soil degradation risk will be composed of indicators of soil compaction, soil organic matter, soil erosion and soil salinisation. In common with other countries' experience, the development of component indicators is well advanced in some areas but still at a relatively conceptual stage for the biodiversity indicator and its species and habitat components. However, with an annual budget of over C\$260 000 (US\$ 169 000), the programme represents a significant commitment to indicator development and use in this sector.

France is an example of a country where work on the broad national environmental indicators covering the whole economy and on specific agri-environmental indicators has advanced at the same time. The Institut Français de l'Environnement (IFEN), which is a public agency reporting to the Ministry of the Environment on scientific and statistical information, has recently produced a report on agricultural and environmental indicators in France (IFEN, 1997a). It is intended to complement a more general report on environmental indicators (IFEN, 1997b). It proposes a set of 49 national indicators designed as a tool for increasing understanding of the relationship between farming and the environment, for quantifying the environmental pressures and benefits attributable to agriculture, and monitoring the effect of action designed to improve the effects of such impacts. Some of the proposed indicators are included in the Appendix, and the main categories used are:

- input use (fertilizers and pesticides);

- natural resources (soils, water, energy);

- atmospheric emissions;

- habitats, biodiversity and landscapes; and

- the role of agriculture in maintaining rural equilibrium (social and environmental).

In the *United Kingdom*, the Ministry of Agriculture is following a similar process. A recent consultation document proposes a list of 35 indicators of sustainable agriculture, most of which are acknowledged to be Driving Force or Response measures (MAFF, 1998). The criteria used for selecting these indicators were that they should be directly relevant to policy, analytically sound, measurable, appropriately aggregated at national level, and representative of public concerns. Again, indicators are grouped into a number of categories, although these are more numerous (12) and more detailed than those used in either the Canadian or the French documents. The aim is to agree a definitive set of indicators in 1999 which specifically relate to agriculture's effects upon the environment, and complement broader UK government work on sustainable development indicators (see the Appendix for a selection of the indicators proposed).

Since 1992, the *Japanese* government has been promoting "sustainable agriculture" by means of improving soil fertility with organic matter, and reducing certain inputs such as chemical fertilizers and pesticides. Two policy approaches are taken: diffusion of information which enhances farmers' environmental awareness, and development and dissemination of technology to mitigate the environmentally negative impacts of agriculture without loss of its productivity. However, agricultural experts and farmers were initially critical that, without some quantitative indication of the extent to which conventional practices had harmful impacts on the environment, and precisely which recommended conservation practices could mitigate the impacts, they could not determine the costs of adopting the recommended agricultural practices.

Since 1993, alongside the OECD work, the Japanese government has been developing agri-environmental indicators and organising work by agricultural research institutes, in order both to contribute to the OECD work and to promote national policies to advance the sustainability of agriculture.

At present, the following indicators are being developed:

- Water resource indicators:

  - Flood prevention

  - Groundwater recharge

  - Water quality

- Land resource indicators:

  - Soil erosion prevention

  - Landslide prevention

  - Soil fertility

- Other indicators:

  - Pesticides

  - Biodiversity

  - Greenhouse gases

  - Landscape.

Although the development of the indicators is still at a preliminary stage, some results have been provided for discussion by the "Investigation Council on Basic Problems concerning Food, Agriculture and Rural Areas", which submitted a report on the orientation of new agricultural policies in September 1998.

When available, the agri-environmental indicators developed by the OECD will be utilised in current research in Japan. For example, research in trends in nitrogen balances in each of the 47 prefectures is underway.

A scientifically sound set of indicators, mainly on bio-physical linkages between agriculture and the environment, will continue to be developed in Japan. Rigorous evidence is essential, given the sensitivities of policy interventions affecting farmers' choice of agricultural practices.

As the agricultural land area in Japan is limited, agri-environmental indicators based on actual measurement, rather than a modelling approach, will be developed. Indicators which can identify impacts at the individual farm level are needed, given that there are 3.5 million farms in Japan, averaging 1.1 hectares per holding, and the impact of each farm on the environment is therefore small.

Other examples of early strategic work by policy makers to develop agri-environmental indicators include *Norway* (Snellingen Bye and Mork, 1998) and *New Zealand* (Ministry for the Environment, 1997).

3.3   National examples of policy evaluation work

Policy evaluation is becoming more widespread and the use of indicators for evaluation of environmental, agricultural and agri-environmental policies is gaining ground. Evaluations of existing operational policies are the most widespread, but there are also appraisals of future policy options where indicators have been used. The indicators range from rather broad spectrum variables, such as land use change, to relatively precise ones, such as some of those concerned with pesticides. One recent example of prior appraisal is provided by *Canada* where a guide to environmental evaluation of agricultural policy options has been produced by the agricultural authorities (Campbell, 1998). The guide includes a series of matrices to be filled in by the person undertaking the evaluation. The basic approach is outlined in Box 8.

Box 8. Assessment of environmental impacts of agriculture policy options in Canada

The following steps are proposed in a guide to agriculture policy evaluation produced by Agriculture and Agri-Food, Canada:

1.   Estimate the potential effects of the policy on the decisions made by producers, e.g. allocation of land, crop selection, nutrient applications, soil conservation, wildlife management, pest management drainage, irrigation, livestock concentration, etc. (12 categories identified).

2.   Estimate whether the management decisions identified in Step 1 may give rise to one or more environmental risks, e.g. soil erosion or compaction, loss of organic matter, water contamination by nutrients, pesticides or bacteria, change in habitat quantity or diversity, air pollution, etc. (13 categories identified).

3.   Try to determine the significance of the environmental risks identified in Step 2 which can be established by four criteria:

— Is the effect of large magnitude, e.g. affecting a large area of land, soil or water?
— Does it affect the amount of a resource that is beyond an established critical threshold of quality or quantity?
— Is the effect long-term or irreversible?
— Does it relate to issues which are sensitive in the wider public debate?

Indicators are provided to help answer these questions, primarily the first two. They include area of land or habitat affected, numbers of contaminated wells or water sources, quantities of nutrient or sediment entering water courses, tolerable erosion rate, agreed water quality standards, etc.

4.   Significant risks are to be summarised and incorporated with significant economic and social impacts in an overall evaluation.

Source: Campbell, 1998.

In the EU, all member States are obliged to implement an agri-environmental incentive scheme for farmers, known as Regulation 2078/92. It is obligatory for the authorities to undertake evaluation of the measures which they have put in place to implement this policy. Although evaluation began rather late in practice in many countries, there has been a major advance over the last 18 months. The European Commission has prepared a non-binding list of "elements for monitoring and evaluating the impact of agri-environmental programmes". This includes general questions about uptake in areas of application and questions about socio-economic, agricultural and environmental impacts. The latter include indicators such as the concentration of nitrates and pesticides in water.

Demand for indicators is also being generated by new legislation, for example where it sets down obligatory environmental standards affecting the agricultural sector (e.g. for water quality), or establishes new processes, such as environmental impact assessment of projects or programmes. In *Denmark*, for example, indicators for nitrate, phosphate and pesticide loading are now used to monitor the effectiveness of water quality legislation, and the area of land under organic production is used as an indicator of the success of the Government's Organic Action Plan (Danish Environment Protection Agency, 1998). Clearly, *Sweden* has laid down a number of targets for the agriculture sector in legislation and policy decisions. For example, it is intended to enrol 600 000 hectares of farmland in agri-environmental schemes, to convert 10 per cent of arable land to ecological forms of production by the year 2000, and to reduce total consumption of inorganic fertilizers by 20 per cent by the end of the century (Government of Sweden, 1996).

There are also examples of indicators being used as part of the process of monitoring the impact of policies, although it can be difficult to relate monitoring results to specific policy measures. In the *United Kingdom*, monitoring of agri-environmental measures has been in place for the past ten years and the techniques used have evolved considerably over that period. In general, there has been a trend away from attempts to measure policy effectiveness using detailed scientific, site-based ecological assessments of change, because these tend to be very costly and are often unable to detect change over a short time-scale or to relate observed changes directly to policy impacts. While systematic assessment of ecological change continues to have an important role, several recent monitoring initiatives have attempted to make more integrated assessments of initial goals and implementation (targets, budgets, uptake), early indications of habitat or feature response to scheme management, and peer assessment of success (involving environmental and agricultural groups). Specific indicators have been devised for several of these elements. The new methods have been selected in order to increase the explanatory power of monitoring, and its ability to generate short term recommendations for on-going scheme improvement. However, these indicators are often only relevant in the context of the particular schemes and geographical locations being considered, for example:

- In certain geographically-defined zones where schemes are in place, indicators may include lengths of traditional field boundary and abundance of particular priority species of flora or fauna, where increases are considered both feasible and beneficial to the environment. Outside these areas, these indicators may be less relevant.

- Indicators may measure the success of particular policies in meeting their own targets and priorities — clearly, these measures are only relevant for schemes which share the same targets.

3.4   Examples of policy targeting

Another important application for indicators is in the targeting of agri-environmental schemes. Many such schemes are open to all farmers who meet certain criteria, or are limited purely by budgetary constraints, with applicants being treated on a first come, first served basis. However, there are examples of schemes where potential participants are subject to an evaluation process, often involving ranking of competing bids. One example is the *Countryside Stewardship Scheme* in England, which uses a scoring system based upon the assessed benefits to wildlife, landscape, historic value and public enjoyment of each proposal, as well as whether the proposal meets particular regional priorities, which are agreed and published annually. Another is the *Conservation Reserve Program* in the United States, participation in which is now subject to a ranking exercise based on an Environmental Benefits Index developed by the United States Department of Agriculture (see Box 9).

3.5   The choice of indicators

Most of the indicators selected by policy makers for their own use fall broadly within the 13 categories identified by the OECD (see OECD, 1997). Examples from several countries are shown in the Appendix. Within these categories, specific indicators often reflect national priorities and preoccupations, such as the water buffering capacity of agricultural fields and infrastructure in Japan. For the most part, this diversity mirrors the variety of purposes required, and where indicators serve similar purposes it is common to find broadly comparable timetables being selected.

Some indicators fall outside the OECD categories. Of these, the most prominent cluster is concerned with energy, with indicators in several countries attempting to capture energy or fossil fuel use within agriculture itself, or within a larger span of the food chain countries. Indicators are also used in some cases to show the level of production of energy crops, which may be measured by area grown or weight of output. Since biomass is a form of renewable energy, production on farmland is regarded as a direct contribution to environmental goals in many countries. The OECD could consider adding in its own work, a category covering energy issues.

4.   Some issues for policy makers

In deciding whether and how to use indicators and which ones to select, policy makers are confronted with a range of practical, technical, conceptual, political and institutional issues. Several of these are discussed in other papers presented in this report but bear brief repetition here.

4.1   Data issues

At this stage of development, data issues are a major concern in most Member countries.

The availability of relevant data on the environment and on agricultural systems varies considerably within the OECD. This reflects several factors, including the size, wealth and historic socio-cultural-political preoccupations of the country and past and present policy priorities. The extent to which earlier environmental and agricultural legislation has led to the collection of data relevant to current policy concerns can also be important. In the EU, for example, the drinking water Directive, agreed in 1980, led to a sharp increase in the monitoring of drinking water in the member States, revealing the extent of nutrient and pesticide concentrations for the first time in some cases (Baldock and Bennett, 1991). Monitoring of key environmental variables does not occur routinely in all OECD countries. Monitoring effort is often driven by legislative requirements, and frequently is not designed to pick out the particular role of agriculture in influencing environmental change.

Box 9. US Department of Agriculture *Environmental Benefits Index* (EBI)

This index is used for ranking bids by landowners to enrol in the Conservation Reserve Program in its most recent form — Sign-Up 16. It is composed of six environmental factors plus a separate cost factor which awards points for low cost bids and those where no government cost-share contribution is required. Bids are compared at a national level and those expected to produce the greatest environmental benefits, based on this system of ranking, are selected. The six environmental factors are:

1. *Wildlife Factor* (0-100 points)
    a. Wildlife Habitat Cover Benefits (0-50 points)
    b. Endangered Species (0-15 points)
    c. Proximity to Water (0, 5 or 10 points)
    d. Adjacent to Protected Areas (0, 5 or 10 points)
    e. Contract Size (0, 2 or 5 points)
    f. Restored Wetland to Upland Percentage (0. 1, 5 or 10 points)

2. *Water Quality Factor* (0 to 100 points)
    a. Location Points
    b. Ground water Quality Benefits
    c. Surface Water Quality Benefits
    d. Wetland Benefit Points

3. *Erosion Factor* (0-100 points)
    The potential for wind or water erosion is measured using an Erodability Index (EI)

4. *Enduring Benefits Factor* (0-50 points)
    This reflects appropriate vegetation or management practices likely to remain beyond the contract period, including tree planting, wetland restoration, etc.

5. *Air Quality Benefits from Reduced Wind Erosion* (0-35 points)
    a. Wind Erosion Impacts (0-25 points)
    b. Wind Erosion Soils List (0 or 5 points)
    c. Air Quality Zones (0 or 5 points)

6. *State or National Conservation Priority Areas (CPAs)* (0 to 25 points)
    The number of points depend on the location of the land proffered, e.g. within national or state designations.

Source: USDA (1997b).

Many state indicators have been built on general national data which is not intended to reveal the particular contribution of agriculture to environmental change. It may require fresh research and investment to identify the particular influence of agriculture and allow more appropriate agri-environmental indicators to be fashioned. Such work can be expensive, particularly in countries with a large land area, such as Canada, and may not be the priority of environmental agencies charged with data collection.

Shortages of data will continue to inhibit the work on indicators in several countries, especially where detailed information is required on environmental change on farmland over time. Precise information about farm management and changing practice is in limited supply in many countries, and this will make it more difficult to develop sophisticated indicators which capture critical aspects of management from an environmental perspective. Cost is a critical factor for most policy makers considering investment in data. It may be difficult to justify a major national survey of environmental factors, such as the state of landscapes, habitats and water quality on farmland, although this may be required to establish a consistent national baseline for the agriculture sector from which future change can be measured.

In a range of countries, information on the rural environment and agri-environmental relationships is being collected more widely and more systematically than in the past. The United States, Japan and EU are examples of this trend. The demand for new data comes from various sources including legislative pressures, growing concern about water pollution and other high profile issues, and an increased appreciation of agriculture's role in managing the rural environment. Furthermore, the increasing number of incentive schemes and other agri-environmental measures in place in OECD countries has led to a new demand or information to clarify the precise origins of and solutions to environmental problems, to identify sensitive and target areas, to establish socio-economic and technical trends, to estimate appropriate payment rates, to measure farmer response and scheme uptake and to evaluate impacts in the short and longer term. Such pressures have led many agriculture agencies and ministries to invest more attention and resources in environmental and policy related data. As well as relying on routine monitoring and specially commissioned scientific and socio-economic research, several ministries have given new responsibilities to executive agencies, made greater use of Geographic Information Systems (GIS), developed questionnaires to elicit responses from farmers and stakeholders, and improved the internal flow of information about the performance of their own policies. All these sources of new data will help to inform the future development of indicators.

There is little doubt that the availability of credible, quantified data on certain issues has influenced the weighting given to these issues in the development of indicators. A clear example is in the frequent use of data on fertilizer inputs (nitrates, phosphates etc.), which can be readily quantified but the precise environmental impact of which may vary considerably in time and space, rather than the use of data which actually measures the polluting effects of various agricultural practices on soils, water and biodiversity. For this reason, some of the more sophisticated indicators for sustainable agriculture avoid proxy indicators of this kind, in favour of more tightly specified measures, e.g. the contribution of agriculture to nitrates in freshwater, rather than nitrate use on farms. However, this sort of indicator can be less transparent to farmers and the public, in that it is less clear how farm management can be altered so as to reduce the impact that has been measured.

A number of OECD Member countries have found it possible to identify quantified indicators where the data is available. For example, concentrations of nutrients in potable water, emissions of certain gases to the atmosphere, and quantities of soil eroded from farmland have proved susceptible to a more quantified approach, although the reliability of the data and its sensitivity to regional and local variations within a country varies substantially.

Quantification has made little progress in other areas, such as biodiversity, although several countries are now investing research effort in trying to extend the reach of quantified indicators. Landscape has been one of the topics attracting greater research interest in recent years, for example, within the EU-funded programme of research on indicators, known as ELISA, Environmental Indicators for Sustainable Agriculture, which began in late 1997. However for some aspects of this work, it is acknowledged that quantification faces conceptual, as well as technical, obstacles (see Section 4.2 below).

It may be questioned whether the development of indicators is driven more by policy requirements or by the availability of appropriate data. Both clearly have a role, alongside external forces such as the work of international organisations. However, when choosing which indicators to select for a practical application, such as policy evaluation, data availability becomes a critical concern. Not only must there be a source of data which is fairly readily available, it should be applicable at the scale required by the policy maker, which may be local, regional or national.

4.2.   Other technical, conceptual and institutional issues

A selection of some of the issues arising in developing, proposing and using indicators may help to illustrate some of the preoccupations of policy makers with respect to agri-environmental indicators. Several are recurrent themes.

It can be difficult to develop meaningful quantitative indicators for certain themes at a national level, because of the diversity of natural and cultural assets and attributes within each country. For example, numerical indices of species diversity at national level are meaningless without reference to spatially-differentiated context information, such as habitat types and natural ranges. Similarly, certain features which contribute to landscape quality and coherence in one area may reduce it elsewhere. National-level quantification is often easier and perhaps more relevant with respect to the basic resources of soils, water and air. The particular difficulties of selecting any standardised, internationally relevant indicators for landscape are well rehearsed in one of the background papers to the OECD Workshop on agri-environmental indicators, which concludes that even if landscapes could be classified according to a standardised international typology, different states and trends in certain possible indicators (land use diversity, field boundaries, landscape features etc.) would have different significance depending upon where they were occurring. In cases such as this, some commentators have concluded that highly composite indicators will be needed to achieve any meaningful international comparisons, e.g. measures of landscape integrity or quality.

In designing and using indicators to measure policy performance, there is a need to distinguish between those which measure simply the policy process itself, and those which detect actual environmental outcomes on the ground. Examples of process indicators are common, and include numbers of farms adopting new kinds of environmental plan, the area of land enrolled into agri-environmental schemes, areas of land designated for environmental protection, and so on. These are helpful as indicators of policy implementation, sometimes regarded as "outputs" but should not be confused with indicators of observed environmental change on the ground. Often these process indicators are used as proxies when more precise measures of environmental outcome are not possible. For example, in its first report on the application of agri-environmental measures in the EU, the European Commission was only able to compare policies by reference to uptake and expenditure figures for the different schemes in each member State, because information on outcomes was not yet available for the majority of schemes (Commission of the European Communities, 1997). In a policy area such as this, where understanding is still far from complete, it should not be assumed that trends in environmental outcomes will mirror those observed in process indicators.

For policy makers working on agri-environmental issues, it is important to be clear about the precise linkage of any indicator to agricultural practice. A number of the UNCSD indicators are not agriculture-specific, for example "releases of nitrogen (N) and phosphorus (P) to marine waters" will include domestic and industrial sources of N and P, as well as farm sources. The French, United Kingdom, United States and Canadian indicators discussed previously attempt to be more specific, relating only to outputs from agriculture, but there are still some areas where sector-specific data are lacking.

The criteria which policy makers in OECD countries apply when selecting agri-environmental indicators appear rather similar. In the recent consultation document by the Ministry of Agriculture in the UK, five criteria are specified — policy relevance, analytical soundness, measurability, appropriate level of aggregation and public resonance (MAFF, 1998). In The Netherlands, where the Ministry of Agriculture is engaged in a project investigating the possibilities of establishing one or more pesticide risk indicators for implementation, nine environmental risk indicators are being evaluated against criteria which include policy relevance, clear relationship with environmental impact, suitability for the determination of long-term trends and for retrospective analysis, and international acceptance in the OECD.

Many of the lists of potential indicators published by international bodies and national governments contain large numbers of indicators. In many cases, there is an explicit intention to reduce these to a smaller number in order to make them clearer and more operational. There is some pressure at international level to slim down the presently wide spectrum of agri-environmental indicators into a small handful, or even one or two synthetic indicators which could be considered alongside other established economic indicators such as the Producer Subsidy Equivalent (PSE). This pressure to produce a small number of powerful and convincing indicators sits uncomfortably with much of the scientific work which underlines the need for precision and an ability to capture issues which may be lost during the process of aggregation. Thus far, there is little evidence of any OECD governments having developed and tested a small group of key agri-environmental indicators at a national level. Most are still considering a range of options. Work on pesticide risk indicators in The Netherlands, referred to above, suggests that the most appropriate single indicator will depend on the purpose for which it is required. Whereas one might be most informative in showing the overall trend and the level of risk, another might be preferable for use by individual farmers.

As agri-environmental indicators make the transition from the research to the implementation stage, new institutional issues arise. Typically, the original impetus for work on environmental indicators came from environmental agencies, research bodies, non-governmental organisations and international organisations. At first, they may have seemed remote or largely irrelevant to agricultural authorities. However, it has been common for agricultural ministries to become more involved when international obligations and legal requirements begin to bear down on the farm sector, and more information about the scale and nature of agriculture's role in different environmental issues is required. In developing agri-environmental indicators, the lead may be taken by agricultural ministries or agencies reporting to them, as in most OECD countries, or, more exceptionally, by environmental agencies such as IFEN in France.

In the majority of cases, it is agricultural authorities which are running the new generation of agri-environmental schemes and consequently they have become involved in the use of indicators for targeting and evaluating these measures. However, there is often more than one authority involved in indicator work and it is not always clear how work on specifically agri-environmental indicators meshes in to other national initiatives. Naturally, different agencies and policy makers have their own objectives and time-scales and this can lead to tensions. Co-operation between agriculture and environmental authorities is obviously desirable in the development and implementation of indicators but both sides must feel a sense of ownership of the indicators chosen if they are to be confident of applying them and defending them in public.

Work on sustainable development and sustainable agriculture brings together social, economic and environmental indicators. At present, the relationship between them is not always clear, and there has not been much debate about the relationship between social and environmental indicators for agriculture in an OECD context. If farm size, for example, is a useful socio-economic indicator, how far can it provide environmental information as well? As governments move away from a strong preoccupation with agricultural output and become more explicit in advocating multi-purpose farming systems, at least in some areas, indicators are likely to find greater application and could become a major input into international policy debates. They will not be confined to environmental issues, and it is already clear that considerations, such as animal welfare and food safety, are gaining political priority in many OECD countries. Agri-environmental indicators are likely to be one element in a widening spectrum.

5.   Concluding remarks

A sizeable number of OECD Member countries have been involved in the development of agri-environmental indicators, particularly over the last five years, during which this area of work has emerged as a subject in its own right. Development has been propelled by a mixture of international and domestic pressures, and work within the OECD has exerted a clear influence which is openly acknowledged in many national publications on the subject. The combination of a growing international debate about agricultural trade, environmental standards, subsidies and policy reform at one level, and proliferating agri-environmental schemes at a national and regional level should maintain the interest of policy makers in the further development of indicators. At present, much of the work has been preparatory, adapting more general environmental indicators to the specific requirements of agriculture, reviewing data and analytical work, identifying potential indicators and experimenting with their use, for example for policy evaluation.

Up to now, there has been more emphasis on state indicators than those representing driving forces or responses. However, this is now changing, with many agriculture ministries particularly concerned with both the latter categories. As different agencies have become involved in developing indicators for their own purposes, there has been some proliferation in the number of institutions involved. In several countries, there may be scope for strengthening institutional linkages and increasing co-operation between different actors in order to avoid potential duplication and confusion. However, it is clear that agricultural ministries need to have a major stake in the development and application of indicators if they are to feel confident of applying them in practice, and this requires a continuing dialogue between agricultural and environmental authorities.

APPENDIX

Examples of proposed agri-environmental indicators for certain OECD Member countries

OECD (OECD 1997)	Indicators from other sources within these categories				
Core Agri-environmental Indicators	UNCSD Agenda 21 (UNCSD, 1996)	UNITED KINGDOM (Department of the Environment, 1996 and MAFF, 1998)	FRANCE (IFEN, 1997a)	CANADA (AAFC, 1998)	OTHER SOURCES[1]
Nutrient use	• Use of fertilizers	• Trends in nitrogen use • Nitrate losses to freshwater • Phosphate losses to freshwater • Proportion of soils at different phosphate levels • Proportion of farmland analysed for phosphate	• Phosphate loading from fertilizers and effluents • Average duration of cover crops, and areas sown • Disposal of crop residues by crop • Nutrient surplus of nitrates • Contribution of agriculture to annual pollution by phosphates	Input use efficiency: • efficiency of fertilizer use • partial nitrogen balance • partial phosphorous balance	
Pesticide use	• Use of agricultural pesticides	• Trends in pesticide use • Quantity of active ingredients used • Area of land treated • Residue levels in food	• Trends in usage of active ingredients in tonnes • Compliance of water intended for human consumption with the EC atrazine standard	Input use efficiency: • efficiency of pesticide use (not yet possible to measure due to lack of national data)	
Water use	• Annual withdrawals of ground and surface water as a percentage of available water • Percentage of arable land irrigated	• Abstraction by use including for spray irrigation • Storage capacity on farm as a proportion of water used • Economic value of irrigated crops	• Contribution of agriculture to water abstractions • Areas irrigated and proportion of holdings equipped to irrigate • Reservoirs for irrigation • Proportion of lakes/ponds used for agriculture by area	Input use efficiency: • irrigation by application system efficiency	

(continued)

Land use and conservation	• Land use change • Changes in land condition • Decentralised local-level natural resource management • Land affected by desertification	• Loss of rural land uses to development • Area of agricultural land lost to hard uses • Area restored to agricultural use from landfill or mineral extraction	• Land in agricultural use • Progress with sustainable development plans, by number • Trends in land reorganisation (remembrement)	• Part-elements of Biodiversity and Farm Management indicators	• At least 5% of flat cropland set aside for wetlands, permanently vegetated strips, etc. (Sweden)
Soil quality	• Area affected by salinisation and waterlogging	• Concentration of organic matter in topsoils • Acidity of topsoils • Concentrations of certain heavy metals in agricultural topsoils in England and Wales • Soil management techniques	• Number and density of severe incidents of agricultural soil erosion	• Risk of Soil Degradation • Addition of cadmium should not exceed its rate of removal • Minimum soil phosphate level	• Humus at least 4% of clay and silt soils, pH at least 6 (Sweden)
Water quality		• Levels of pesticides in rivers and in groundwater • Number of wildlife "incidents" (water pollution events)	• Delineation of nitrate vulnerable zones • The rate of quality objectives defined for monitored sections of rivers being met • Concentrations of oxidisable matter, nitrates and phosphorous in freshwater and eutrophication • Pesticide concentrations in water in three regions	• Risk of Water Contamination	• The number of fish kills from agricultural sources (Ireland) • Serious pollution from agricultural sources per km of river (Ireland)
Greenhouse gases		• Emissions of methane from agriculture • Emissions of nitrous oxides from agriculture	• Emissions of carbon dioxide, methane and nitrous oxide from agriculture[2]	• Agro-ecosystem Greenhouse Gas Balance	

(continued)

Biodiversity	• Threatened species as a percentage of total native species	• Native species which are threatened (using IUCN criteria)[3] • Number of British breeding bird species increasing or declining in population, and distribution by broad habitat type • Populations of key farmland birds • Plant diversity in semi-improved grasslands, hedgerows and streamsides • Numbers of mammal species and population size in Britain • Proportion of British butterfly species whose distribution has changed, by landscape type	• Number of threatened vertebrate and plant species (using IUCN criteria)[3] • Number of threatened breeds of domestic animals (donkeys, cattle, horses, pigs, sheep) • Number of conservation measures implemented for rare crop breeds	• Agro-ecosystem Biodiversity Change — species component
Wildlife habitats	• Protected area as a percentage of total area • Protected forest area as a percentage of total • Releases of nitrogen and phosphorous to coastal waters	• Change in total areas of chalk grassland in England • Changes in the numbers and size of chalk grassland sites in Dorset • Estimates of the numbers of lakes and ponds in Great Britain • Area of semi-natural grassland • Area of cereal field margins under environmental management • Area of agricultural land under commitment to environmental conservation	• Proportion of agricultural land uses in designated zones of floral and faunal interest • Trends in wetland areas of France (stable, degraded) • Area of agricultural land in biosphere reserves and nature reserves • Area of land in agri-environmental schemes, by scheme	• Agro-ecosystem Biodiversity Change — habitats component • Agricultural drainage activities expressed as km of ditches per year (Iceland) • Change in area of wetlands (Iceland) • Change in area of protected wetlands (Iceland)

(continued)

Landscape	• Length of hedgerows and walls	• Occurrence of hedges and trees and permanent pasture on farms • Area of agricultural land within national and regional parks • Proportion of air photograph squares dominated by agriculture, or with some agriculture present • Importance of agriculture in urban fringe areas • Age classes of farm buildings (built heritage on farms), and new construction • Growth in areas of woodland	• Distribution and size of isolated cultivated plots (Norway) • Distribution and size of uncultivated areas with an emphasis on their significance to the natural heritage (Norway)
Farm management	• Adoption of environmental management systems by farmers • Use of machinery and techniques which reduce pollution • Soil management practices • Percentage of farmland in organic production or under conversion	• Implementation of codes of good fertilizer practice • Implementation of waste management standards for buildings • Area and proportion of land in organic production by crop	Farm resource management: • soil cover management • land management practices • inputs management including nutrients, pesticides and energy
Farm financial resources	• Producer Subsidy Equivalent • Proportion of CAP expenditure on environmental payments		

(*continued*)

Socio-cultural issues	• Population change in mountain areas • Welfare of mountain populations • Agricultural education	• Rural unemployment	• Importance of agriculture in less-favoured areas and mountains • Agribusiness/food sector employment in rural areas • Young farmers installed • Number of holdings and proportion which are full-time • Agricultural employment	
Energy[4]		• Area planted with energy crops	• Use of energy in farm business • Use of domestic energy on farms • Area planted with energy crops	• Management of energy within inputs management indicator

Notes:

1. Sources for "other" countries:
   Iceland: Brunvoll (1997);
   Ireland: Honohan (ed.) (1997);
   Norway: Information from the Norwegian Ministry of Agriculture;
   Sweden: Swedish Environmental Protection Agency (1997).
2. French indicators also include the production of other air pollutants causing acid rain, from agriculture — sulphur dioxide and "photochemicals".
3. IUCN — World Conservation Union.
4. This indicator is not part of the current OECD core agri-environmental indicators.

Source: Author (1998).

BIBLIOGRAPHY

AGRICULTURE AND AGRI-FOOD CANADA [AAFC] (1998), *Agri-Environmental Indicator Project — Summary of activities in fiscal year 1997-8*, Ottawa.

BALDOCK, D. and G. BENNETT (1991), *Agriculture and the Polluter Pays Principle: A Study of Six EC Countries*, Institute for European Environmental Policy, London.

BRUNVOLL, F. (ed.) (1997), "Indicators of the State of the Environment in the Nordic Countries", *TemaNord*.

CAMPBELL, I. (1998), *Guide to the Environmental Evaluation of Agricultural Policies and Programs*, Environment Bureau Policy Branch, Agriculture and Agri-Food Canada, Ottawa.

COMMISSION OF THE EUROPEAN COMMUNITIES (1997), *Report on the application of Council Regulation (EEC) No. 2078/92 on agricultural production methods compatible with the requirements of the protection of the environment and the maintenance of the countryside*, COM(97)620, Brussels.

DANISH ENVIRONMENT PROTECTION AGENCY (1998), *Danish Action Programme to Prevent and Reduce Nitrate Losses from Agriculture,* internal paper, Copenhagen.

DEPARTMENT OF ENVIRONMENT, UNITED KINGDOM (1996), *Indicators of Sustainable Development for the United Kingdom*, HMSO, London.

DUMANSKI, J., S. GAMEDA and C. PIERI (1998), *Indicators of Land Quality and Sustainable Land Management*: *An Annotated Bibliography*, a joint publication of the World Bank and Agriculture and Agri-Food Canada, World Bank, Washington, D.C.

EUROSTAT (1998), *Environmental Indicators included in the Environmental Pressure Indices Project* (see EUROSTAT world-wide web Internet page).

GOVERNMENT OF SWEDEN (1996), *Country Profile. Implementation of Agenda 21. Review of Progress Made since the United Nations' Conference on Environment and Development, 1992*, submission to the UN's Commission on Sustainable Development by the Swedish Government, Stockholm.

HEIMLICH, R.E. (1995), "Environmental Indicators for US Agriculture" in S. Batie (ed.), *Developing Indicators for Environmental Sustainability: The Nuts and Bolts*, Special Report (SR) 89, Proceedings of the Resource Policy Consortium Symposium, Washington, D.C.

HOLTEN-ANDERSEN, J., H. PAABY, N. CHRISTENSEN, M. WIER and F. MOLLER ANDERSEN (1995), *Recommendations on Integrated Environmental Assessment*, Report submitted to the European Environment Agency by the National Environmental Research Institute (NERI), Denmark.

HONOHAN, P. (ed.) (1997), *EU Structural Funds in Ireland: A Mid-term Evaluation of the CSF 1994-99*, The Economic and Social Research Institute, Dublin.

IFEN (1997a), *Agriculture et environnement : les indicateurs*, édition 1997-1998, Institut français de l'environnement (IFEN), Paris.

IFEN (1997b), *Environmental Performance Indicators in France 1996-1997*, Institut français de l'environnement (IFEN), Paris.

MINISTRY FOR THE ENVIRONMENT, NEW ZEALAND (1997), "Environmental Performance Indicators: proposals for air, fresh water, and land", *Signposts for Sustainability*, October, Wellington.

MINISTRY OF AGRICULTURE, FISHERIES AND FOOD, UNITED KINGDOM (1998), *Development of a Set of Indicators for Sustainable Agriculture in the United Kingdom: A Consultation Document*, June, HMSO, London.

OECD (1997), *Environmental Indicators for Agriculture*, Paris.

OFFICE OF THE AUDITOR GENERAL, CANADA (1997), *Integrated Economic/Environmental Modelling* — a presentation by Agriculture and Agri-Food Canada, October, Ottawa.

SNELLINGEN BYE, A. and K. MORK (1998), *Resultatkontrolljordbruk 1998*, Statistics Norway.

SOMBROEK, W.G. (1997), "Land resources evaluation and the role of land-related indicators" in *Land Quality Indicators and their Use in Sustainable Agriculture and Rural Development*, Proceedings of the Workshop organised by the Land and Water Development Division, FAO Agriculture Department and the Research, Extension and Training Division, FAO Sustainable Development Department, FAO Land and Water Bulletin 5, 25-26 January 1996, World Bank/UNEP/UNDP/FAO, Rome.

STANNERS, D. and P. BOURDEAU (1995), *Europe's Environment: the Dobrís Assessment* (with Statistical Assessment), European Environment Agency, Copenhagen.

SWEDISH ENVIRONMENTAL PROTECTION AGENCY [SEPA] (1997), *The Agriculture of the Future*, Summary of Final Report on a Systems Study of Sustainable Agriculture, Stockholm.

UN COMMISSION ON SUSTAINABLE DEVELOPMENT [UNCSD] (1996), *Indicators of Sustainable Development Framework and Methodologies*, United Nations, New York, United States.

US DEPARTMENT OF AGRICULTURE [USDA] (1997a), *Agricultural Resources and Environmental Indicators, 1996-7*, Economic Research Service, Agricultural Handbook Number 712, July, Washington, D.C.

US DEPARTMENT OF AGRICULTURE [USDA] (1997b), *Environmental Benefits Index*, Economic Research Service, Fact Sheet, October, Washington, D.C.

WORLD BANK (1997a), *World Development Indicators*, Washington, D.C.

WORLD BANK (1997b), *Expanding the Measure of Wealth Indicators of Environmentally Sustainable Development*, Washington, D.C.

PART IV:

OFFICIAL STATEMENTS

OPENING ADDRESS TO THE WORKSHOP

by
Elliot Morley,
Member of Parliament, Countryside Minister,
Ministry of Agriculture, Food and Fisheries,
London, United Kingdom

I am grateful to Peter Smith for introducing me and for chairing this opening plenary session and setting the scene for the workshop.

I am delighted to welcome you to this workshop on behalf of the United Kingdom and to see representatives from so many countries and organisations. Of course, the workshop would not be taking place without additional funding from other countries and I would like to thank Austria, Canada, Finland, Japan, Norway, Spain, Sweden, Switzerland and the United States for their financial support.

I hope that this workshop will be as successful as the previous ones held in Madrid on agriculture and forestry, Helsinki on the environmental benefits of agriculture and Athens on the sustainable management of water in agriculture.

My department and the UK Department of the Environment, Transport and the Regions have worked closely in organising this workshop. I am pleased that a senior official from the Department of the Environment — Mrs Sophia Lambert, Director, Wildlife and Countryside — will be chairing the final session.

We have chosen the location for the workshop very carefully. I'm sure you will all be aware by now that York is an exceptionally attractive and historic city, with excellent facilities and connections to the rest of the UK and abroad. These were some of the factors which led to the Ministry of Agriculture to relocate part of its headquarters functions here and to build a new site for the Central Science Laboratory which many of you visited yesterday.

Equally important is York's central position in one of this country's major agricultural areas. The region might be said to encapsulate the diversity of this country's agriculture industry within a small area — highly productive arable, pig and poultry farming in the Vale of York contrasting with sheep farming and management of red grouse in the hills. And North Yorkshire has some of the finest agricultural landscapes in England, containing a rich legacy of habitats and historic and archaeological features created and maintained by farming activities.

A number of programmes are in place to help farmers preserve and enhance the environmental interest of the region. Some of these are operated by the Ministry of Agriculture such as our Countryside Stewardship Scheme or by other bodies such as the North York Moors National Park.

But not all initiatives involve Governments paying farmers to carry out environmental good works. For example, the agriculture industry has established an organisation called *LEAF — Linking Environment and Farming* — which promotes integrated crop management. LEAF has a network of demonstration farms, including two in Yorkshire, which provide farmers with practical demonstrations of how they can use integrated crop management to protect the environment and maintain their profitability.

The study visit on Thursday has been designed to show you the United Kingdom's approach to promoting sustainable agriculture and I hope that you will enjoy it.

This workshop is being held at a critical time. Work on developing agri-environmental indicators in the OECD Joint Working Party (JWP) on agriculture and the environment has been going on since 1993 and it is important that it is now brought to fruition. The development of indicators will contribute to the wider work of OECD.

The time taken to advance this work is perhaps not surprising. As we have found in the United Kingdom, devising indicators for agriculture is not straightforward since its impacts are complicated, diverse and liable to substantial local variation. And agriculture is one of those activities, as is forestry, that generate beneficial as well as harmful effects on the environment.

It is difficult enough to produce indicators to suit all the circumstances of an individual country — let alone the 29 OECD Member countries. So it is to the credit of the JWP that you have already been able to make so much progress. We have certainly found your work a stimulus to our own efforts to develop agri-environmental indicators.

This brings me to the **first of the two points** which I want to make this morning. One of the main uses of indicators is, rightly, as a tool to help policy makers in Government assess the impacts of agriculture and to help identify the best policy approaches. But they should also mean something to the people they concern.

In the UK we see the primary purpose of the JWP's indicators as being for policy analysis but they also have the potential to encourage more sustainable behaviour in the farming industry. After all, sustainable agriculture is not just a matter for Governments; it can only be realised through the activities of individual farmers, landowners and managers. This means that indicators need to be rooted in the realities of agricultural production if they are to be regarded as credible and influence day to day management decisions on farms.

Equally, it is important that indicators reflect the expectations of the public about the environmental performance of agriculture and its contribution to sustainable development.

I know that the JWP has had access to a wide range of agronomic, scientific and other technical expertise in developing its indicators. Indeed, I think that this is one of its strongest features of its work. But I would urge you to ensure that the indicators reflect the concern of the agriculture industry, Non-Governmental Organisations (NGOs) and the wider public as you move closer to finalising them.

In the United Kingdom, for example, the Ministry of Agriculture recently issued a public consultation document containing proposals to establish a set of indicators for sustainable agriculture.

The consultation document will be followed by a seminar later this year so that we can listen to the views of a range of NGOs. Another initiative, being led by the Department of the Environment, is to identify a small number of headline indicators which capture the key sustainability issues in a more populist way than we have been able to achieve so far.

I recognise that it is not easy to establish consensus. Agriculture seems to be one of those sectors which provokes many differing assessments of its environmental impacts and contribution to sustainable development. However, the process of trying to achieve consensus can at least help to clarify the key issues and to identify areas of common understanding. With this in mind, I particularly welcome delegates representing international farming and environmental organisations at this workshop.

The **second point** which I would make to guide you in your discussions is that the indicators should relate closely to policies aimed at promoting sustainable development, rather than the pursuit of numbers for their own sake or the development of indicators only of interest to the specialist.

I was particularly struck by one of the thoughts in Professor David Pearce's paper which he will present to you following my address, cautioning against developing indicators that are primarily data-generated rather than issue-generated. I would add that one of the strengths of the OECD work is to start from identifying the policy issue to be addressed and then search for the appropriate indicators.

In the United Kingdom we have similarly sought to articulate the policy issues first and then to identify the indicators to illustrate them. For example, the starting point of our consultation document on indicators for sustainable agriculture is a vision of what a sustainable agriculture might look like.

Another initiative, which is being led by the Department of the Environment colleagues, is to review the United Kingdom's sustainable development strategy issued in 1994 and the set of indicators of sustainable development published in 1996.

These indicators will be linked explicitly to the issues set out in the review of the strategy so that we have a transparent measure of our progress towards sustainable development.

I should also say that, by sustainable development, we mean an approach which seeks to integrate environmental, economic and social considerations and which seeks to achieve four key objectives: social progress; effective protection of the environment; prudent use of natural resources; and economic growth and employment.

I know that this is also the vision of sustainable development in the OECD and you might consider during this workshop whether the JWP's indicators should be called "indicators for sustainable agriculture" rather than "environmental indicators for agriculture".

I am aware that the JWP's indicators are firmly rooted in its programme of work to analyse the environmental impacts of agriculture and agricultural policies. We strongly support the OECD's efforts to identify policies which can achieve environmentally and economically sustainable agriculture at minimal resource cost to the economy whilst avoiding trade distortions.

Although sustainable development is of great international interest, there is as yet no internationally accepted framework within which to develop measures of sustainability. The OECD has an important leading role in developing such a framework and I hope that this workshop will help to bring it one step closer.

I wish you all a successful workshop which I am sure will lead to sound, practical recommendations relevant to policy makers. I look forward to meeting many of you during this morning's coffee break.

OUTCOME OF THE WORKSHOP IN THE CONTEXT OF OECD WORK ON SUSTAINABLE DEVELOPMENT

by
Gérard Viatte,
Director for Food, Agriculture and Fisheries, OECD,
Paris, France

I am most grateful to Sophia Lambert for her introduction and agreeing to take on the important task of chairing this concluding session of the Workshop.

This Workshop has been a great success on two counts. Firstly, our hosts have made us very welcome in this beautiful part of the United Kingdom, and on behalf of the OECD Secretariat I wish to express my sincere thanks to the Minister, Elliot Morley, to Dudley Coates and Sophia Lambert and their staff, for their efficient co-ordination and organisation of this Workshop and the study visit.

I should also add that co-operation with the UK and the Secretariat in organising the Workshop has worked extremely well. May I also take this opportunity to thank my own staff for their very hard work.

Secondly, constructive recommendations have emerged from this Workshop, which reflects the dedication of all participants in preparing papers and contributing to the discussion. Within our work programme, the activity on agri-environmental indicators receives a high priority and has to be carried out, like all our activities, in a very cost-efficient manner and with a sense of selectivity. The Workshop recommendations will help to define the future development of this activity.

We should also remember that the Workshop has provided an excellent basis to develop the dialogue between policy makers, scientists and researchers, governments, farmers and environmentalists.

I would like to make four points that expand upon the Minister Morley's opening address.

First, in the more detailed discussion on agri-environmental indicators it is important not to lose sight of the broader context of the "twin pillars" of agricultural policy reform and sustainable development.

In this regard, early 1998 has been an important period for OECD, with a meeting in March of OECD Agriculture Ministers and in April of Environment Ministers. Also an initiative has been launched by the OECD Secretary General on Sustainable Development.

At the OECD Agriculture Ministerial meeting, Ministers outlined their shared goals for agriculture and agreed to a set of policy principles and operational criteria to guide agricultural policy reform.

The Ministers agreed *inter alia* that governments should work to ensure that the agro-food sector contributes to the sustainable management of natural resources and the quality of the environment. In this respect actions are needed so that farmers take both environmental costs and benefits into account in their decisions. This requires information and indicators that reveal the magnitude and trends in the effects agriculture is having on the environment.

In fact the former UK Agriculture Minister, Dr. Jack Cunningham, emphasised the importance of developing agri-environmental indicators during the OECD Agriculture Ministerial meeting.

Indicators have the potential to better inform the policy decision-making process. They can help to underpin the operational criteria for sound policies outlined by Ministers for Agriculture: by providing transparent and targeted information on agri-environmental policy issues, reflecting the diversity of agri-environmental situations, responding to changing policy priorities, shedding light on the implications of policy choices.

In deepening work on integrating environmental concerns into key sectors such as agriculture, OECD Environment Ministers recommended that the OECD should further develop a set of "robust indicators to measure progress towards sustainable development" and to intensify efforts to upgrade the extent and quality of environmental data and indicators.

The work on agri-environmental indicators will also provide one of the building blocks in developing a set of OECD sustainable development indicators, within the OECD horizontal project on sustainable development. The report from the Sustainable Development project, is aimed at providing a policy strategy to help achieve the economic, social and environmental dimensions of sustainable development.

These three dimensions are recognised in the OECD indicators to measure sustainable agriculture, taking into account the need to produce sufficient and safe food, while addressing environmental concerns and meeting social goals.

In this context, I was particularly interested in Elliot Morley's suggestion that we change the name from "agri-environmental indicators" to "indicators of sustainable agriculture". It is an idea that we should follow up in the OECD.

This brings me to my **second point**. OECD has always been proud of its rigorous and objective analysis supported by robust data.

The OECD also has a core competence in providing a forum to achieve an international consensus on policy-relevant, comparable indicator definitions and methodologies. An excellent and well-known example is OECD work on the internationally recognised Producer and Consumer Subsidy Equivalents as a measure of agricultural support levels.

The Workshop has, I believe, successfully risen to the challenge in recommending policy-relevant indicators for analysis of agri-environmental issues. But we must continue to be selective in developing the most important indicators. The Workshop has started that process, and the questionnaire that many countries completed prior to the Workshop, as well as the criteria proposed during the discussions, gave us a very useful broad indication of those priorities, as well as the feasibility to develop selected indicators.

In this context, OECD Ministers of Agriculture in their Communiqué stressed that there is a need for further research, a better understanding of current scientific knowledge, and better supply of information to users to tackle such issues as environmental performance. As with all data, the indicators will need to be refined and improved as policy and public priorities alter, the environmental, economic and social situation evolves, and scientific knowledge improves.

On a personal note, in a speech I made in the UK almost a decade ago to the Agricultural Economics Society, I said "It is important that agricultural research contributes to clarifying the still often inconsistent scientific data and providing guidance for policy makers and farmers." That is as true today as in 1990.

My **third point** concerns co-operation. OECD Member countries, and other players in the indicator field will play a vital role in the future success of the work. This is already evident from the work underway on agri-environmental indicators in nearly all countries and highlighted by the papers and resources countries have committed to the achievements of this Workshop.

I am also very encouraged that the effort to develop indicators is being extended to a wide range of stakeholders, a point made by Elliot Morley. OECD Agriculture Ministers emphasised the need for the OECD to involve key players in promoting an active policy dialogue. Indeed, I particularly value the contribution to this forum by the representatives from farming organisations, environmental interest groups, and international organisations, such as the FAO and UNEP.

It would be remiss of me not to stress that the role of Member countries and other players in the future development and prioritisation of agri-environmental indicators will also be crucial given the limited resources available in the OECD Secretariat and in Member countries for this activity. But the results so far are testimony to what can be achieved from such co-operation.

OECD is increasingly co-operating with non-Member countries, and much of what we have accomplished in this Workshop can also be of value to these countries. Most of the environmental issues of importance to the OECD agricultural sector are also of global significance. The indicators we are developing, such as on soil quality and water use, should have a wider application to other countries.

My **fourth point** looks forward to the role of indicators in the future policy dialogue. A key theme which runs through a number of the papers at this Workshop is that indicators should be driven by policy issues rather than data availability. Indicators must not been seen as an end in themselves. They must serve as an essential tool for better informing the policy dialogue and decision making. And they must be understandable to the public: indicators will need to be interpreted.

This approach is central to that adopted by OECD with the identification of policy issues followed by indicator development.

The indicator work will in the future provide a vital element to help quantitative agri-environmental policy analysis and monitoring work that is an integral part of the activities of the OECD's Joint Working Party and its parent Committees.

The development of indicators coincides with a critical turning point in the domestic and international agri-environmental policy area. This is illustrated by the agricultural policy reforms currently underway in most OECD countries, which often include an environmental component, the Kyoto Protocol commitments to reduce greenhouse gas emissions, and the forthcoming negotiations under the auspices of the World Trade Organisation.

Finally, I would stress that agricultural and environmental policies cannot be considered in isolation. An integrated approach is required. We all need to build bridges and linkages. A good example for this has been set by the active participation of many experts at this Workshop from both Agriculture and Environment Ministries. And I am confident that, together, we have significantly contributed to that process of integration.

Keynes once said that economists "must examine the present, in the light of the past, for the purposes of the future". Although Keynes also famously said that "in the long run we are all dead", in the meantime we need good indicators to help us chart the future — not least to provide a better environment for our children and grandchildren!

CLOSING REMARKS FOR THE WORKSHOP

by
Dudley Coates,
Head of Environment Group,
Ministry of Agriculture, Food and Fisheries,
London, United Kingdom

I am pleased that it falls to me — on behalf of the UK — to make some closing remarks about what has clearly been a very successful, if demanding, workshop.

I should like to thank all the consultants, chairs, rapporteurs, presenters and discussants — as well as the delegates — for contributing to its success. Thanks also goes to Austria, Canada, Finland, Japan, Norway, Spain, Sweden, Switzerland and the United States for their additional financial contributions, without which the workshop would not have been possible.

Last but not least, we are all indebted to the OECD Secretariat, in particular Wilfrid Legg (Head of the Policies and Environment Division, OECD Agriculture Directorate) and his team, and to my colleagues at the UK Ministry of Agriculture and the Department of the Environment, Transport and the Regions for their hard work. The agenda for this workshop was particularly complex and required a great deal of preparation to ensure that everything ran as smoothly as it did. And I should not forget the efforts of all those who made the tour of the Central Science Laboratory and the study visit so enjoyable.

Turning to the outcome of the workshop, as our Countryside Minister, Mr Morley, emphasised at the start, sustainable development is by its nature an international issue. It requires an internationally agreed framework for measuring whether countries are becoming more sustainable.

I believe that this workshop has reinforced the leading role of the OECD in developing such a framework. The OECD is particularly well suited to this because it has been able to combine its tradition of rigorous economic analysis with an objective examination of countries' policies and their implications for sustainable development.

Of course, individual countries will want to have their own indicators as well which reflect their own traditions and circumstances. As Mr Morley said, ownership of a problem by those most closely involved with it can be the first step to finding a solution, and indicators are likely to be effective at changing policies and behaviour only if they are closely linked and relevant to the problem in hand.

It is also a reality that the OECD has limited resources for this work — as Gérard Viatte has pointed out — so it will need to husband its resources and concentrate on its strengths. I suggest that OECD can best help countries to develop their own indicators by providing an overarching framework which —

- identifies the common ground between OECD countries, whose circumstances and conditions can vary widely;

- develops agreed methodologies for calculating indicators;

- highlights trends across OECD countries in the major policy areas; and

- uses the information generated by indicators as one element in analysing policies across the OECD.

In the UK, we have always regarded the policy driven approach of the OECD as a particular strength. Indeed, I think it is accepted by the OECD Joint Working Party (JWP) that the purpose of the indicators is to provide a tool to facilitate its wider work programme on agriculture and the environment. So it is important that we do not to lose sight of the overall objective of this work and this is an opportune moment to remind you of the Joint Working Party's remit, which is —

> *"to identify ways in which governments might design and implement policies and promote market solutions to achieve environmentally and economically sustainable agriculture at minimal resource cost to the economy and with least trade distortions."*

I know that the Joint Working Party's indicators have been long in gestation and that there have been criticisms that the work is taking too long to come to fruition. So, it is a crcdit to you all that this workshop has made significant steps to progress the Joint Working Party's indicators.

In particular, I have been impressed by the extent to which the workshop has managed to integrate the development of specific indicators in the break-out sessions with the cross-cutting issues discussed in the plenary sessions.

Thank you once again for the tremendous efforts you have all made to ensure that this has been a highly successful workshop. Now it remains for the OECD Secretariat, in co-operation with OECD countries, to put the various recommendations into a workable and measurable form. I wish you all a safe journey home and hope we have the opportunity to meet again in the future.

ANNEX

LIST OF RECOMMENDED OECD AGRI-ENVIRONMENTAL INDICATORS

This Annex provides a detailed list of OECD agri-environmental indicators and their definitions, recommended for both "short- and long-term development", summarised into five tables, as follows:

Annex Table 1 —
Contextual Indicators
1. Land
2. Population
3. Farm Structures

Annex Table 2
3. Water Quality
4. Water Use
5. Soil Quality
6. Land Conservation

Annex Table 3
7. Biodiversity
8. Wildlife Habitat
9. Landscape

Annex Table 4
10. Farm Management
    — Farm Management Capacity
    — On-farm Management Practices
11. Farm Financial Resources
12. Socio-cultural Issues (Rural Viability)

Annex Table 5*
13. Nutrient Use
14. Pesticide Use
15. Greenhouse Gases

---

* These three areas where not discussed at the York Workshop, but OECD work is underway to develop indicators that address these areas, see OECD (1999), *Agricultural Policies in OECD Countries: Monitoring and Evaluation 1999*, Chapter IV, Volume I, Paris, France.

Annex Table 1. List of recommended contextual indicators

Land	Population	Farm structures
Agricultural Land use and Land Cover Changes 1. Changes between the share of land in agriculture and other uses 2. Changes in the share of agricultural land cover type	**Number of full-time farmers** Changes in the number of full-time farmers	**Number and type of farms** Changes in farm types and numbers

Annex Table 2. List of recommended indicators proposed for the water, soil and land areas

Water Quality	Water Use	Soil Quality	Land Conservation
Nitrate concentration in water in agricultural vulnerable areas The proportion of ground and surface water, in agricultural vulnerable areas, above a reference level of nitrate concentration (NO_3 mg/l)	**Water use intensity** The proportion of water resources subject to diversion for agricultural use	**Risk of soil erosion by water** The agricultural area subject to water erosion (i.e. the area for which there is a risk of degradation by water erosion above a certain reference level)	**Water buffering capacity** The quantity of water that can be stored over a short period, *in* the agricultural soil, as well as *on* agricultural land where applicable (e.g. flood storage basins) and *by* agricultural irrigation and drainage facilities

Water Quality	Water Use	Soil Quality	Land Conservation
Phosphorus concentration in water in agricultural vulnerable areas The proportion of surface water bodies, in agricultural vulnerable areas, above a reference level of phosphorus concentration (P$_{total}$ mg/l)	**Water stress** The proportion of rivers subject to diversion for irrigation without Defined Minimum Reference Flows	**Risk of soil erosion by wind** The agricultural area subject to wind erosion (i.e. the area for which there is a risk of degradation by wind erosion above a certain reference level)	
Risk of water contamination by nitrogen The area of agricultural land potentially vulnerable to water contamination by nitrogen			
Risk of water contamination by pesticides The area of agricultural land potentially vulnerable to water contamination by pesticides	*Water use technical efficiency* For selected irrigated crops, the mass of agricultural produce (tonnes) per unit of the volume of irrigation water consumed, the latter being the volume of water, in megalitres, diverted or extracted for irrigation less return flows	*Inherent soil quality* Agricultural areas where there is a mismatch between the soil capability as indicated by the index of inherent soil quality and the actual or impending land use	*Off-farm sediment flow* The quantity of soil sediments delivered to off-farm areas from agricultural soil erosion

Annex Table 2 (continued). List of recommended indicators proposed for the water, soil and land areas

Water Quality	Water Use	Soil Quality	Land Conservation
	Water use economic efficiency For all irrigated crops, the monetary value of agricultural production per unit of irrigation water volume consumed, the latter being the volume of water, in megalitres, diverted or extracted for irrigation less return flows		
	Policy and management response to water stress Indicates potential economic distortions in the use of water caused by underpricing, free access or government intervention in the management of irrigation water, in particular, in countries or regions with a high intensity of water use		

Indicators in **bold** are for short-term development and those in *italics* are for medium- to long-term development.

Annex Table 3. List of recommended indicators proposed for the biodiversity, wildlife habitat and landscape areas

Biodiversity	Wildlife Habitat	Landscape
Genetic diversity of domesticated livestock and crops 1. Change in the sum of all recognised and utilised varieties of domesticated livestock and crops 2. Change in the share of different livestock and crop varieties in the total population or in total livestock and crop production	**Intensively farmed agricultural habitats** The share of each crop in the agricultural area.	**Land characteristics of agricultural landscape** 1. Natural features, covering, for example, the land's slope, elevation, soil type, etc. 2. Environmental appearance, including the landscape ecosystems and habitat types 3. Land type features, including changes in agricultural land use and land cover type
Wildlife species diversity related to agriculture A. Quality 1. Appropriate key species indicators for each agro-ecosystem 2. Key threatening processes that can damage agricultural production activity 3. Proportion of semi-natural and uncultivated natural habitats on agricultural land B. Quantity 4. The extent of changes in the agricultural area and type of land cover (this indicator would draw from the wildlife habitat and land use/cover indicators)	**Semi-natural agricultural habitats** The share of the agricultural area covered by semi-natural agricultural habitats	**Cultural features of agricultural landscape** Key indicative cultural features

Annex Table 3 (continued). List of recommended indicators proposed for the biodiversity, wildlife habitat and landscape areas

Biodiversity	Wildlife Habitat	Landscape
	Uncultivated natural habitats	**Management functions of agricultural landscape**
	1. Area of wetland transformed into agricultural area 2. Area of aquatic ecosystems transformed into agricultural area 3. Area of natural forest transformed into agricultural area 4. Area of agriculture re-converted into aquatic ecosystems	The share of agricultural land under public and private commitment to landscape maintenance and enhancement
Change in numbers of endangered species related to agro-ecosystems	*Habitat heterogeneity (average size of habitats)*	*Landscape typologies*
Impacts on biodiversity of different farm practices and systems	*Habitat variability (number of habitat types per monitoring area)*	*Monetary valuation of societal landscape preferences (from public surveys)*
Effects on biodiversity caused by off-farm soil sediment flow	*Impact on habitat of different farm practices and systems*	

Indicators in **bold** are for short-term development and those in *italics* are for medium- to long-term development.

Annex Table 4. List of recommended indicators proposed for the farm management, farm financial and socio-cultural (rural viability) areas

Farm Management		Farm Financial Resources	Socio-cultural (Rural Viability)
Farm management capacity	**On-farm management practices**		
Standards for environmental farm management practices Number of established national and/or sub-national environmental farm management standards, regulations, codes of practice, etc.	**Matrix of environmental farm management practices** The matrix includes an issue substructure (nutrients, soil, pesticides, water, etc.) and specified management practices under each, with countries reporting on the level of adoption or "actual" use of those practices most relevant to their specific national and regional situations. The focus in the short term should be on measuring specific management practices; both the share of farms (or land area) using the practice and its implementation	**Public and private agri-environmental expenditure** Public and private expenditure on agri-environmental goods, services and conservation (both investment and current expenditure)	**Agricultural income** Share of agricultural income in relation to total income of rural households

Annex Table 4 (continued). List of recommended indicators proposed for the farm management, farm financial and socio-cultural (rural viability) areas

Farm Management		Farm Financial Resources	Socio-cultural (Rural Viability)
Farm management capacity	**On-farm management practices**		
Expenditure on agri-environmental research Expenditure on agri-environmental research as a percentage of total agricultural research expenditure		**Farm financial equilibrium** The equilibrium between the net farm operating profit after tax (i.e. farm monetary receipts), and the cost of capital (i.e. financial costs to the farm)	**Entry of new farmers into agriculture** Number of farmers, according to age and gender, entering the agricultural sector
Educational level of farmers Average educational attainment of farmers, presented as the share of farmers attaining different levels of education or years of education			

Annex Table 4 (continued). List of recommended indicators proposed for the farm management, farm financial and socio-cultural (rural viability) areas

Farm Management		Farm Financial Resources	Socio-cultural (Rural Viability)
Farm management capacity	**On-farm management practices**		
Ratio of agricultural advisers Number of public and private agricultural advisers trained in environmental management practices per farmer	*Implementation index* The Implementation Index could be used to measure the extent to which environmental farm management practices are actually used by farmers. It would be a way to express the results of the matrix of environmental farm management practices (see above) in a comprehensive manner for a given country	*Adjusted farm financial equilibrium* Adjusting farm financial resources for changes in natural resource depletion and pollution, for example, soil erosion and nutrient soil surface balance	*Social capital in agricultural and rural communities* The strength of social institutions and formal/informal networks, voluntary organisations, etc., in agricultural and rural communities

Indicators in **bold** are for short-term development and those in *italics* are for medium- to long-term development.

Annex Table 5. List of indicators for nutrients, pesticides and greenhouse gases

Nutrient Use	Pesticide Use	Greenhouse Gases
Nutrient balances (soil surface balances of nitrogen and phosphorous)	**Index of pesticide use (active ingredients)**	**Gross agricultural emissions (methane, nitrous oxide and carbon dioxide)**
Farm gate nutrient balance *Nutrient use efficiency (technical and economic)*	*Pesticide use efficiency (technical and economic)* *Pesticide risk indicators*	*Agriculture's contribution to renewable energy (biomass production)* *Net emissions of carbon dioxide from agricultural soils* *Economic efficiency of agricultural greenhouse gas emissions*

Indicators in **bold** are for short-term development and those in *italics* are for medium- to long-term development. These indicator areas were not discussed at the York Workshop, but are included among the areas for which the OECD is developing agri-environmental indicators.

WORKSHOP AGENDA AND MAIN CONTRIBUTORS

York, United Kingdom, 22-25 September 1998

PLENARY SESSION 1 —
OPENING OF THE WORKSHOP AND OVERVIEW OF THE ISSUES

Chair: Peter Smith (United States Department of Agriculture)
Opening of the Workshop: Elliot Morley (Member of Parliament, Countryside Minister, Ministry of Agriculture, Food and Fisheries, United Kingdom)

Measuring Sustainable Development: Implications for Agri-environmental Indicators
David Pearce (University College, London, United Kingdom)

Discussant: Allan Haines (Ministry of Environment, Australia)

Taking Stock of the OECD Work on Agri-environmental Indicators
Kevin Parris (OECD Secretariat)

Cross-cutting Issues in Developing Agri-environmental Indicators
Andrew Moxey (University of Newcastle-upon-Tyne, United Kingdom)

Discussant: Jukka Peltola (Agricultural Economics Research Institute, Finland)

BREAKOUT GROUPS —
CHAIRS, RAPPORTEURS, PRESENTERS AND DISCUSSANTS

Group 1: Water Quality / Water Use / Soil quality / Land Conservation

Chair: Joseph Racapé (Ministry of Environment, France)
Rapporteur: Chris Doyle (Scottish Agricultural College, United Kingdom)
OECD Secretariat: Gérard Bonnis and Seiichi Yokoi

Water Quality
Presenters: Eiko Lubbe (Federal Ministry of Food, Agriculture and Forestry, Germany)
Jessper S. Schou (Danish Institute of Agricultural and Fisheries Economics, Denmark)
Discussant: Ted Huffman (Agriculture and Agri-Food Canada, Canada)

Water Use

Presenter: Leslie Russell (Department of Primary Industries and Energy, Australia)
Discussant: José Antonio Ortiz Fernandez-Urrutia (Ministry of Agriculture, Fisheries and Food, Spain)

Soil Quality

Presenter: Richard Arnold (United States Department of Agriculture)
Discussant: Winifried Blum (University of Resource Sciences, Austria)

Land Conservation

Presenter: Katsuyuki Minami (Ministry of Agriculture, Forestry and Fisheries, Japan)
Discussant: Tomasz Stuczynski (Ministry of Agriculture and Food Economy, Poland)

Group 2: Biodiversity / Wildlife Habitats / Landscape

Chair: David Purcell (Permanent Delegation of Australia to the OECD, Paris)
Rapporteur: Gerry Hamersley (English Nature, United Kingdom)
OECD Secretariat: Kevin Parris

Biodiversity

Presenter: Ben Ten Brink (Ministry of Environment, The Netherlands)
Discussant: Jorge Soberón (CONABIO, Mexico)

Wildlife Habitats

Presenter: Daniel Zürcher (Swiss Federal Office for Environment, Forests and Landscape)
Discussant: Steve Brady (United States Department of Agriculture)

Landscape

Presenter: Dirk Wascher (European Centre for Nature Conservation, The Netherlands)
Discussant: Göran Blom (Swedish Environmental Protection Agency, Sweden)

Group 3: Farm Management / Farm Financial Resources / Socio-cultural Issues

Chair: Gabriella Dånmark (Ministry of Agriculture, Norway)
Rapporteur: Philip Lowe (University of Newcastle-upon-Tyne, United Kingdom)
OECD Secretariat: Outi Honkatukia and Morvarid Bagherzadeh

Farm Management

Presenter: Robert Koroluk (Agriculture and Agri-Food Canada, Canada)
Discussant: Ian Davidson (Ministry of Agriculture, Fisheries and Food, United Kingdom)

Farm Financial Resources

Presenter: Nicola Shadbolt (Massey University, New Zealand)
Discussant: Judith Hausheer (Federal Research Station for Agricultural Economics and Engineering, Switzerland)

Socio-cultural Issues

Presenter: Frank Clearfield (United States Department of Agriculture)
Discussant: Bernard Dechambre (Ministry of Agriculture and Fisheries, France)

Study Visit organised by the United Kingdom authorities

PLENARY SESSION 2 —
DISCUSSION OF OUTCOME FROM BREAKOUT GROUPS

Chair: Terence McRae (Agriculture and Agri-Food Canada, Canada)

PLENARY SESSION 3 —
ROLE OF AGRI-ENVIRONMENTAL INDICATORS IN POLICY ANALYSIS

Chair: Kunio Tsubota (Ministry of Agriculture, Forestry and Fisheries, Japan)

Developing and Using Agri-environmental Indicators for Policy Purposes: OECD Country Experiences
David Baldock (Institute for European Environment Policy, United Kingdom)

Discussant: Kevin Steel (Ministry of Agriculture and Forestry, New Zealand)

Using Agri-environmental Indicators to Assess Environmental Performance
Professor Paul Thomassin (Chair, McGill University, Canada)

Discussant: Ralph Heimlich (United States Department of Agriculture)

PLENARY SESSION 4 —
SUMMARY AND RECOMMENDATIONS OF THE WORKSHOP

Chair: Sophia Lambert (Department of the Environment, Transport and the Regions, United Kingdom)

Presentation of the Key Recommendations of the Workshop
Wilfrid Legg (OECD Secretariat)

Outcome of the Workshop in the Context of OECD Work on Sustainable Development
Gérard Viatte (Director, Directorate for Food, Agriculture and Fisheries, OECD)

Concluding Comments
Dudley Coates (Ministry of Agriculture, Food and Fisheries, United Kingdom)

— Closure of the Workshop —

LIST OF PARTICIPANTS

AUSTRALIA Allan HAINES, Environment Australia, Canberra
Ray JEFFERY, Department of Primary Industries and Energy (DPIE), Canberra
David PURCELL, Permanent Delegation of Australia to the OECD, Paris
Mary PATTON, Australian Bureau of Statistics, Canberra
Leslie RUSSELL, Department of Primary Industries and Energy (DPIE), Canberra

AUSTRIA Winfried E.H. BLUM, University of Agricultural Sciences, Vienna
Franz GOLTL, Osterreichisches Statistisches Zentralamt, Vienna
Johannes SCHIMA, Prasidentenkonferenz der Landwirtschaftskammern
Österreichs, Vienna
Helmuth WALTER, Federal Ministry of Agriculture and Forestry, Vienna
Gerhard ZETHNER, Federal Environment Agency, Federal Ministry of
Environment, Youth and Family, Vienna

BELGIUM Erik BOMANS, Soil Service of Belgium, Heverlee
Jacques CORNET, Environmental Protection Agency (DGNE), Namur
Johan HEYMAN, Ministère de l'Agriculture, Relations Internationales, Brussels
Ludo VANONGEVAL, Soil Service of Belgium, Heverlee

CANADA Ted HUFFMAN, Research Branch, Agriculture and Agri-Food Canada, Ottawa
Robert KOROLUK, Economic and Policy Analysis Directorate, Policy Branch,
Agriculture and Agri-Food Canada, Ottawa
Terence McRAE, Environment Bureau, Agriculture and Agri-Food Canada, Ottawa
Ted WEINS, Prairie Farm Rehabilitation Administration, (PRFA), Regina

CZECH REPUBLIC Jiri HRBEK, Division of Agriculture, Czech Statistical Office, Prague

Pavel ZAVAZAL, Division of Agriculture, Czech Statistical Office, Prague

DENMARK Anne Vibeke JACOBSEN, Danmarks Statistik, Copenhagen
Søren Søndergaard KJÆR, Danish Environmental Protection Agency, Copenhagen
Dorrit KRABBE, Ministry of Food, Agriculture and Fisheries, Copenhagen
Arne KYLLINGSBAEK, Danish Institute of Plant and Soil Science, Tjele
Jesper Sølver SCHOU, Danish Institute of Agricultural and Fisheries Economics,
Valby

FINLAND Into KEKKONEN, Ministry of Environment, Helsinki
Jukka PELTOLA, Agricultural Economics Research Institute, Helsinki

FRANCE	Bernard DECHAMBRE, Ministère de l'Agriculture et de la Pêche, Paris Jean-Marie DEVILLARD, Ministère de l'Agriculture et de la Pêche, Paris Joseph RACAPÉ, Ministère de l'Aménagement du Territoire et de l'environnement, Paris
GERMANY	Theo AUGUSTIN, Federal Ministry of Agriculture and Forestry, Bonn Eiko LUBBE, Federal Ministry of Food, Agriculture and Forestry, Bonn Hiltrud NIEBERG, Federal Agricultural Research Centre, Institute of Farm Economics, Braunschweig Hans-Peter PIORR, Center for Agricultural Landscape and Land Use Research (ZALF), Müncheberg Astrid THYSSEN, Federal Ministry for the Environment, Nature Conservation and Nuclear Safety, Bonn
GREECE	Liza PANAGIOTOPOULOU, Ktimatologio S.A.(Hellenic Cadastre S.A.), Athens Marlena TIKOF, Ministry of Agriculture, Athens Giannoula VRANAKI, Ministry of Agriculture, Athens
HUNGARY	István FÉSÜS, Plant Protection and Agri-Environmental Management Department, Ministry of Agriculture, Budapest
IRELAND	Dan GAHAN, Department of Agriculture, Food and Forestry, Dublin
ITALY	Andrea FAIS, INEA, Rome Paolo BAZZOFFI, Istituto Sperimentale per lo Studio e la Difesa del Suolo, Ministry of Agricultural Policies, Florence Rosa FRANCAVIGLIA, Istituto Sperimentale per la Nutrizione delle Piante, Ministry of Agricultural Policies, Rome Marcello MASTRORILLI, Istituto Sperimentale Agronomico, Bari Andrea POVELLATO, INEA, Legnaro
JAPAN	Keiichi ISHII, Ministry of Agriculture, Forestry and Fisheries, Tokyo Katsuyuki MINAMI, Department of Research Planning and Coordination, National Institute of Agro-Environmental Sciences, Tsukuba Masamichi SAIGO, Ministry of Agriculture, Forestry and Fisheries, Tokyo Hiroshi SEINO, Division of Environmental Planning, National Institute of Agro-Environmental Science, Tsukuba Kunio TSUBOTA, Research Management and Information, Japan International Research Centre for Agricultural Sciences, Tsukuba Tetsuo USHIKUSA, Permanent Delegation of Japan to the OECD, Paris Makoto YOKOHARI, Institute of Policy and Planning Sciences, University of Tsukuba, Tsukuba

KOREA	Hee-Yeol KIM, Ministry of Agriculture and Forestry, Seoul
	Sang-Jae LEE, Ministry of Agriculture and Forestry, Seoul
	Moo-Eon PARK, National Institute of Agricultural Science and Technology (RDA), Suweon
	Jae-Sung SHIN, National Institute of Agricultural Science and Technology (RDA), Suweon
MEXICO	Arturo CALDERON, Mexican Agricultural Office, Mission of Mexico to the EC, Ministry of Agriculture, Livestock Rural Development, SAGAR, Brussels
	Germán GONZALEZ-DAVILA, Permanent Delegation of Mexico to the OECD, Paris
	Jose SARUKHAN KERMEZ, CONABIO, National Commission for Biodiversity, Mexico
	Jorge SOBERÓN, CONABIO, Institute of Ecology, National University, Mexico City
	Carlos TOLEDO-MANZUR, SEMARNAP, Ministry of Environment, Mexico City
THE NETHER-LANDS	Floor BROUWER, Agricultural Economics Research Institute (LEI-LO), The Hague
	Tjeerd DE GROOT, Ministry of Agriculture, Nature Management and Fisheries, The Hague
	Jan HUININK, National Reference Centre for Agriculture, Blede
	Ed van KLINK, National Reference Centre for Agriculture, Blede
	Ben TEN BRINK, National Institute for Public Health and Environmental Protection, Bilthoven
NEW ZEALAND	Wayne BETTJEMAN, Ministry for the Environment, Wellington
	Nicola SHADBOLT, Massey University, Palmerston North
	Kevin STEEL, Ministry of Agriculture and Forestry, Wellington
NORWAY	Gabriella DÅNMARK, Ministry of Agriculture, Oslo
	Frode LYSSANDTRAE, Ministry of Agriculture, Oslo
	Henrik MATHIESEN, Norwegian Institute of Land Inventory, Oslo
	Akse ØSTEBROT, Directorate of Nature Management, Tronoheim
	Anette SØRAAS, Norwegian Ministry of Agriculture, Oslo
POLAND	Jerzy KOZLOWSKI, Ministry of Agriculture and Food Economy, Warsaw
	Anna LIRO, Ministry of Agriculture and Food Economy, Warsaw
	Grazyna SAS, Ministry of Agriculture and Food Economy, Warsaw
	Tomasz STUCZYNSKI, Institute of Soil Science and Plant Cultivation (IUNG), Pulawy
SPAIN	Angel BARBERO MARTIN, Ministerio de Agricultura, Pesca y Alimentacion, Madrid
	Fernando ESTIRADO-GOMEZ, Ministerio de Agricultura, Pesca y Alimentacion, Madrid

SPAIN	Manuel LOPEZ-ARIAS, Ministerio de Agricultura, Pesca y Alimentacion, Madrid José Ramón LÓPEZ PARDO, Permanent Delegation of Spain to the OECD, Paris Santiago LOPEZ-PINEIRO, Ministerio De Medio Ambiente, Madrid Jose Antonio ORTIZ FERNANDEZ-URRUTIA, Ministerio de Agricultura, Pesca y Alimentacion, Madrid
SWEDEN	Göran BLOM, Swedish Environmental Protection Agency, Stockholm Solveig DANELL, Statistics Sweden, Stockholm Thomas HAGMAN, Swedish Ministry of Agriculture, Stockholm Olof JOHANSSON, Swedish Board of Agriculture, Jönköping Eva LINDHOLM, Swedish Ministry of Agriculture, Stockholm Ingrid SVEDINGER, Swedish Ministry of Agriculture, Stockholm Johan WAHLANDER, Swedish Board of Agriculture, Jönköping
SWITZER-LAND	Marianne ALTORFER, Swiss Federal Agency for Environment, Forests and Landscape (BUWAL), Berne Brigitte DECRAUSAZ, Swiss Federal Office for Agriculture, Berne Johannes DETTWILER, Swiss Federal Agency for Environment, Forests and Landscape (BUWAL), Berne Judith HAUSHEER, Swiss Federal Research Station for Agricultural Economics and Engineering, Tänikon Hans-Jörg LEHMANN, Swiss Federal Office for Agriculture, Berne Daniel ZÜRCHER, Swiss Federal Agency for Environment, Forests and Landscape (BUWAL), Berne
UNITED KINGDOM	Elliot MORLEY, Member of Parliament, Countryside Minister, Ministry of Agriculture, Food and Fisheries, London Dudley COATES, Head of Environment Group, Ministry of Agriculture, Food and Fisheries, London Sophia LAMBERT, Director, Wildlife and Countryside, Department of the Environment, Transport and the Regions, London Nikolaj BOCK, Ministry of Agriculture, Fisheries and Food, London Gary BECKWITH, Ministry of Agriculture, Fisheries and Food, London Julie COLLINS, Countryside Commission, Cheltenham Ian DAVIDSON, Ministry of Agriculture, Fisheries and Food, London Steven GLEAVE, Ministry of Agriculture, Fisheries and Food, London Michael HARRISON, Ministry of Agriculture, Fisheries and Food, London Gwaine HOGG, Ministry of Agriculture, Fisheries and Food, London Andrew HOWARD, Ministry of Agriculture, Fisheries and Food, London David JONES, Ministry of Agriculture, Fisheries and Food, London Michael MARKS, Farming and Rural Conservation Agency, London Stephen REEVES, Department of the Environment, Transport and the Regions, London Andrew STOTT, Department of Environment, Transport and Regions, Bristol Richard SMITH, Environment Agency, Exeter Roger WATTS, HM Treasury, London

UNITED STATES	Richard ARNOLD, United States Department of Agriculture, Washington, D.C.
	Stephen J. BRADY, United States Department of Agriculture, Fort Collins
	Frank B. CLEARFIELD, United States Department of Agriculture, Washington, D.C.
	Ralph E. HEIMLICH, United States Department of Agriculture, Washington, D.C.
	Ronald L. MARLOW, United States Department of Agriculture, Washington, D.C.
	Peter F. SMITH, United States Department of Agriculture, Washington, D.C.

COMMISSION OF THE EUROPEAN COMMUNITIES

Anna BARNETT, Directorate General XI - Environment and Agriculture, Brussels
Hans-Christian BEAUMOND, Directorate General VI - Agriculture, Brussels
Marina GRASSART, Directorate General VI - Agriculture, Brussels
Valéry MORARD, Directorate General VI - Agriculture, Brussels
Eric WILLEMS, Directorate General VI - Agriculture, Brussels
Dirk WASCHER, European Centre for Nature Conservation, Tilburg, The Netherlands

EUROSTAT	Rosemary MONTGOMERY, EUROSTAT, Luxembourg
	Maria PAU VALL, EUROSTAT, Luxembourg
	Claude VIDAL, EUROSTAT, Luxembourg
	Gerd EIDEN, EUROSTAT, Luxembourg

EUROPEAN ENVIRONMENT AGENCY (EEA)

Eileen BUTTLE, Chippenham, Wiltshire, United Kingdom

INTERNATIONAL GOVERNMENTAL ORGANISATIONS

FOOD AND AGRICULTURE ORGANISATION (FAO)

Jeff TSCHIRLEY, Rome, Italy

UNITED NATIONS ENVIRONMENT PROGRAMME (UNEP)

Arthur DAHL, UNEP Earthwatch, Geneva, Switzerland

INTERNATIONAL NON-GOVERNMENTAL ORGANISATIONS

BIRDLIFE INTERNATIONAL

Matthew RAYMENT, Bedfordshire, United Kingdom

EUROPEAN CONFEDERATION OF AGRICULTURE (ECA)

Hermann SCHULTES, Vienna, Austria

INTERNATIONAL FEDERATION OF AGRICULTURAL PRODUCERS (IFAP)

Andrew CLARK, London, United Kingdom

THE WORLD CONSERVATION UNION (IUCN)

Riccardo SIMONCINI, Florence, Italy

WORLD-WIDE FUND FOR NATURE (WWF)

Hilmar von MUENCHHAUSEN, Frankfurt, Germany

CONSULTANTS David BALDOCK, Institute for European Environmental Policy, London, United Kingdom
Andrew MOXEY, Department of Agricultural Economics and Food Marketing, University of Newcastle-upon-Tyne, Newcastle, United Kingdom
David PEARCE, University College, London, United Kingdom
Paul J. THOMASSIN, McGill University, Quebec, Canada

RAPPORTEURS Christopher J. DOYLE, Scottish Agricultural College, Auchincruive, United Kingdom
Gerry HAMERSLEY, English Nature, Peterborough, United Kingdom
Philip LOWE, Centre for Rural Economy, University of Newcastle-upon-Tyne, Newcastle, United Kingdom

OECD SECRETARIAT **Directorate for Food, Agriculture and Fisheries**
Gérard VIATTE, Director
Wilfrid LEGG, Head of Policies and Environment Division
Françoise BÉNICOURT, Assistant to Head of Division
Kevin PARRIS, Principal Administrator
Seiichi YOKOI, Principal Administrator
Morvarid BAGHERZADEH, Administrator
Gérard BONNIS, Administrator
Outi HONKATUKIA, Administrator
Jane KYNASTON, Conference Organiser

Environment Directorate
Jan Horst KEPPLER, Administrator
Jeannie RICHARDS, Administrator
Teresa COSTA-PEREIRA, Administrator

Territorial Development Service
Heino von MEYER, Expert (territorial indicators)

OECD PUBLICATIONS, 2, rue André-Pascal, 75775 PARIS CEDEX 16
PRINTED IN FRANCE
(51 1999 05 1 P) ISBN 92-64-17041-3 – No. 50735 1999